PLUMBERS HANDBOOK

by Joseph P. Almond, Sr.

THEODORE AUDEL & CO.
a division of
HOWARD W. SAMS & CO., INC.
4300 West 62nd Street
Indianapolis, Indiana 46268

Copyright © 1969, 1971, 1973, 1976 and 1979 by
Joseph P. Almond, Sr.

FIFTH EDITION
FIRST PRINTING—1979

All rights reserved. Reproduction or use, without express permission, of editorial or pictorial content, in any manner, is prohibited. No patent liability is assumed with respect to the use of the information contained herein. While every precaution has been taken in the preparation of this book, the publisher assumes no responsibility for errors or omissions. Neither is any liability assumed for damages resulting from use of the information contained herein.

International Standard Book Number: 0-672-23339-8
Library of Congress Catalog Card Number: 79-64811

Printed in the United States of America

PREFACE

This book was written for the sole purpose of assisting in the training of apprentices in the plumbing and pipefitting trades.

This book contains timesaving illustrations plus many tips and shortcuts leading to accurate installations. Such subjects as hub and no-hub cast-iron fittings, copper drainage fittings, plastic pipe, silver brazing, soft soldering, and lead work are included in this book.

Also featured are illustrations on Vents and Venting, the Sovent System, Septic Tanks, Water Heaters, the Sloan Flush Valve, Solar System Water Heater, Oasis Water Cooler, plumbing tools, lots of related math in English and Metric, and working drawings of every imaginable situation or example in plumbing important enough to illustrate.

In the classroom or on the job, it is information at your fingertips—simplified and condensed.

I wish to give recognition to the National Bureau of Standards and its Office of Weights and Measures; the American National Metric Council; the Cast-Iron Soil Pipe Institute; Copper Development Assoc., Inc.; American Standard; the Compressed Gas Association; the Ridge Tool Co.; the Sloan Valve Co.; the A. O. Smith Corp., Consumer Products Div.; NIBCO Inc.; Lead Industries Assoc.; Plastics Pipe Institute; and Oasis Water Coolers, for their contributions towards this book's success. I also wish to thank Dennis Bellville along with my many other friends and associates who have assisted in making this book possible.

I am also very proud to dedicate this revised edition to my devoted wife Suzanne Almond and my precious son Joseph P. Almond, III, both of whom I love very much.

Joseph P. Almond, Sr.

CONTENTS

Metric Information	9
Copper Sovent System	16
Copper Drainage Fittings and Specifications	32
Vents and Venting Illustrated	42
Example in Venting	
Vent and Drainage System—Residence	
Venting Wall Hung Water Closets	
Introduction to the CISPI	55
Members of the CISPI	
Cast-Iron Soil Pipe Fittings and Specifications	57
Minimum Offsets Using Cast Iron Hub Fittings	
Cast-Iron No-Hub Pipe Fittings and Specifications	97
Minimum Offsets Using Cast Iron No-Hub Fitting	
Silver Brazing and Soft Soldering	137
Illustrations on Brazing and Soldering	143
Lead Work with Illustrations	153
Lead Joining Work	
Roughing and Repair Information	163
Lavatories	
Water Closet Flush Tank Illustrated	
How to Replace a Ball Cock or Float Valve	

How to Install a Typical Tub Trip-Waste
   and Overflow
How to Install a Pop-Up Drain
Bathroom Illustration, Layout, and Isometric
Washing Machine Rough-In
Stall Urinal
Repair Water Faucets and Valves

Working Drawings 180
Fresh Air System
Connections Between Heater and Storage Tank
Grease Trap
Acid Diluting Tank
Sand Trap
Electric Cellar Drain
Trailer Connection Rough-In
½ S or P-Trap and Its Parts
Garbage Disposal—Residential
Tub Rough-In
Pressure Reducing Valve
A Bathroom Illustrated in Minimum Space
Gas Hot Water Heater
Septic Tank

Formulas 197
    Rolling Offsets
    Square Root
    Finding Diagonal of a Square
    Right Triangle
    Running Pipe or Tubing Parallel Using Offsets
        and Maintaining Uniform Spread Throughout
    45° Wye and 1/6 Bend in Offset
    Determining Capacity in Gals. and Liters
        of a Tank
    Table of Constants for Calculating
        Offset Measurements
    Converting Decimal Parts of a Foot
        to Inches and Millimeters
    Fahrenheit and Celsius Scales

Symbols for Plumbing Fixtures 210

Oxyacetylene Basic Safety Measures 212

General Information 215
    Plumbing Defined
    Purpose of Venting
    Safety Reminders
    Knots Commonly Used

Typical Hoisting Signals
Compression Tank
Forming Various Angles by use of a Six Foot Rule
Boiling Points of Water at Various Pressures
Above Atmospheric—in Metric and English

Plastic Pipe and Fittings 236

Lead and Oakum Joints 258

Solar System Water Heater Illustrated 261

Tips for the Plumber—Beginner 267

Sloan Royal Flush Valve Illustrated 271

Wall Hung Water Cooler Illustrated 277

Plumbing Tools Illustrated 281

METRIC INFORMATION HELPFUL TO THE PIPING INDUSTRY

I wish to thank the National Bureau of Standards and its Office of Weights and Measures, along with the American National Metric Council, both of Washington, D.C., for providing me with the necessary conversion factors and other related information needed for the "SI" metrication of the Plumbers Handbook.

The International System of Units (SI) is a modernized version of the metric system established by international agreement.

METRIC — ABBREVIATIONS
(Used in Plumbers Handbook)

ANMC	American National Metric Council
ANSI	American National Standards Institute
NBS	National Bureau of Standards
SI	International System of Units
m	meter
km	kilometer
cm	centimeter
mm	millimeter
kg	kilogram
g	gram
mg	milligram
l	liter
ml	milliliter
N·m	newton-meter
°C	degrees celsius
kPa	kilopascal
J	joule
inHg	inches of mercury
Pa	pascal

METRIC INFORMATION
Length

1 meter =	39.37 inches (English)
1 meter =	1000 millimeters
1 meter =	100 centimeters
1 meter =	10 decimeters
1 kilometer =	.625 miles (English)
1.609 kilometers =	1 mile (English)
25.4 millimeters =	1 inch (English)
2.54 centimeters =	1 inch (English)
304.8 millimeters =	1 foot (English)
1 millimeter =	.03937 inches
1 centimeter =	.3937 inches
1 decimeter =	3.937 inches

"Metric"	(Volume—Liquid)	"English"
3.7854 liters		1 gallon
.946 liters		1 quart
.473 liters		1 pint
1 liter		.264 gals. or 1.05668 qts.
1 liter		33.814 ounces
29.576 milliliters		1 fluid ounce
236.584 milliliters		1 cup

Note: 1 liter contains 1000 milliliters

METRIC INFORMATION
Weight (Mass)

1 kilogram =	2.204623 lbs.
453.592 grams =	1 kilogram
1 gram =	.035 ounces
28.349 grams =	1 ounce
28,349 milligrams =	1 ounce
1 gram =	1000 milligrams
1 kilogram =	1,000,000 milligrams
1 kilogram also =	1000 grams
.02831 kilograms =	1 ounce

Note:
1 lb. =	453,592.37 milligrams
1 lb. =	453.59237 grams
1 lb. =	453592 kilograms

1 metric ton weighs 1000 kilograms or 2204.623 lbs.

"Metric"   "English"

2.59 sq. kilometers =	1 sq. mile
.093 sq. meters	1 sq. foot
6.451 sq. centimeters =	1 sq. inch
.765 cu. meters =	1 cu. yd.
.028316 cu. meters	1cu . foot
16.387 cu. centimeters =	1 cu. inch
1 cubic meter =	35.3146 cu. ft.
929.03 sq. centimeters =	1 sq. ft.

Note:
10,000 sq. centimeters =	1 sq. meter
1 cubic meter =	1,000,000 cu. centimeters or 1000 cu. decimeters
10.2 centimeters of water =	1 kPa of pressure
51 centimeters of water =	5 kPa
1 meter of water =	9.8 kPa
1 cubic foot contains	28,316.846522 cu. centimeters

1 centimeter column of mercury (0°C.)=1.3332239 kPa pressure.

Generally speaking, 6 cm of mercury = 8 kPa pressure.

One cubic meter of air weighs 1.214 kilograms (kg).

Atmospheric pressure of 101.3 kPa will balance or support a column of mercury 76 cm high.

Note: When you know "inches of mercury" (inHg) multiply same by 3.386389 to find the number of kilopascals (kPa).

Note: Inch of water (at 39.2°F.) 4°C.—multiply by 2.49082 to find kPa.

Note: 1 ml of water has a mass of 1 gram.

Note: 1 foot lb. = 1.3558 newton-meter (N·m) (bending moment of toque).
  1 inch lb. = 1.1298 newton-meter (N·m) (bending moment of torque)

Note: 1 British Thermal Unit (mean) = 1.05587 joule.
  1 BTU (International Table) = 1.055056 joule.

Example: 144 BTU's = 152 joule (J)
  Use: Conversion Factor: 1.056
  or 1 BTU = 1.056 joule

Note: To determine: degrees Fahrenheit when Celsius is given: °F. = 1.8C. + 32 or 1.8 × C. + 32
  To determine: degrees Celsius when Fahrenheit is given: °C. = F. − 32 ÷ 1.8 or F. − 32 then ÷ by 1.8

Note: (Acceptable Forms)   (Not Acceptable)
  12 to 20°C.   20° C.
  12°C. to 20°C.   12° to 20°C.

METRIC INFORMATION

Kilopascal (kPa) is the unit recommended for fluid pressure for almost all fields of use, such as barometric pressure, gas pressure, tire pressure, and water pressure.

Note: Atmospheric pressure is: 101 kPa Metric, and 14.7 PSI English 6.894757 kPa = 1 PSI.

To find head pressure in decimeters when pressure is given in kilopascal (kPa) divide pressure by .9794.

To find pressure in kPa of a column of water given in decimeters—multiply decimeters by .9794.

To find head in meters when pressure is given in kilopascal —divide pressure by 9.794.

To find pressure in kPa of a column of water given in meters—multiply meters by 9.794.

Example:

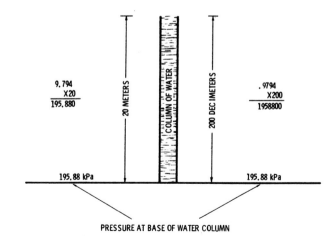

PRESSURE AT BASE OF WATER COLUMN

METRIC INFORMATION

English (inches)	Metric Equiv. (millimeters)
1"	25.4 mm
2"	50.8 mm
3"	76.2 mm
4"	101.6 mm
5"	127 mm
6"	152.4 mm
7"	177.8 mm
8"	203.2 mm
9"	228.6 mm
10"	254 mm
11"	279.4 mm
12"	304.8 mm
"Feet"	
2'	609.6 mm
3'	914.4 mm
4'	1219.2 mm
5'	1524 mm
6'	1828.8 mm
7'	2133.6 mm
8'	2438.4 mm
9'	2743.2 mm
10'	3048 mm
20'	6096 mm

English "Parts of an Inch"	Metric Equiv. "Millimeters"	
1/32"	.79375	(.80)
1/16"	1.5875	(1.6)
1/8"	3.175	(3.2)
3/16"	4.7625	(4.8)
1/4"	6.35	(6.4)
5/16"	7.9375	(7.9)
3/8"	9.525	(9.5)
7/16"	11.1125	(11.1)
1/2"	12.7	(12.7)
9/16"	14.2875	(14.3)
5/8"	15.875	(15.9)
11/16"	17.4625	(17.5)
3/4"	19.05	(19.1)
13/16"	20.6375	(20.6)
7/8"	22.225	(22.2)
15/16"	23.8175	(23.8)
1"	25.4	(25.4)

Note: In most cases it is best to round off to the nearest millimeter. Thus,
17.4625 would be: 17 mm
20.6375 would be: 21 mm

Note: Round off to nearest millimeter. Thus,
4" = 102 mm
3" = 76 mm
2" = 51 mm and so on.

Metric Information

Inches and Parts of An Inch	Centimeters
1/16"	.15875
1/8"	.3175
1/4"	.635
3/8"	.9525
1/2"	1.27
5/8"	1.5875
3/4"	1.905
7/8"	2.2225
1"	2.54
1-1/4"	3.175
1-1/2"	3.81
2"	5.08
2½"	6.35
3"	7.62
4"	10.16
5"	12.7
6"	15.24
7"	17.78
8"	20.32
9°	22.86
10"	25.4
11°	27.94
12"	30.48

Inches and Feet (inches)	Meters
1"	.0254
2"	.0508
3"	.0762
4"	.1016
5"	.127
6"	.1524
7"	.1778
8"	.2032
9"	.2286
10"	.254
11"	.2794
12"	.3048
"Feet"	
1-1/4'	.381
1-1/2'	.4572
2'	.6096
2-1/2'	.762
3'	.9144
4'	1.2192
5'	1.524
6'	1.8288
10'	3.048
25'	7.62
50'	15.24
100'	30.48

COPPER SOVENT

I wish to thank the Copper Development Association Inc., 405 Lexington Ave., New York, N. Y. 10017, for providing the following illustrations and data on the "Copper Sovent System". I am very proud to include this latest advance in the plumbing industry, finely engineered, and well tested.

COPPER SOVENT

"The Single-Stack Plumbing System"

A look at the design, service, experience, background, and important features of this advanced system.

This system was invented in 1959 by Fritz Sommer of Bern, Switzerland.

The "Copper Sovent" single-stack plumbing system is an engineered drainage system developed to improve and simplify soil, waste and vent plumbing in multi-story buildings.

The basic design rules illustrated herein are based on experience gained in the design and construction of hundreds of "Sovent Systems" serving thousands of living units; not to mention the extensive experimental work conducted on the 10 story plumbing test tower.

Copper Sovent

The first "Sovent" installation was made in 1961 in Bern, Switzerland. Eight years later 15,000 apartment units were installed in 200 buildings, up to thirty stories in height.

Through the "Copper Development Association Inc." extensive tests were carried out on the instrumented test tower. Following the successful completion of these early tests the system was brought to America.

The first installation was the "Habitat Apartments" constructed for Expo '67 in Montreal; the next, the "Uniment" structure in Richmond, California. In 1970, two large apartment buildings, and a 10-story office building using "Sovent" were begun.

The great potential of the "Sovent" drainage system lies ahead.

The guidelines presented are intended to be useful to plumbing system design engineers, architects, building owners and mechanics doing the installation.

Each individual "Sovent" system must be designed to meet the conditions under which it will operate, and engineering judgment is required in applying the basic design rules presented here to specific buildings.

"Copper Development Association Inc." will be pleased to review "Sovent" system designs to help assure that the design principles are followed.

The copper "Sovent" plumbing system has four parts:

1. A copper DWV stack.
2. A "Sovent" aerator fitting at each floor level.
3. Copper DWV horizontal branches.
4. A "Sovent" deaerator fitting at the base of the stack.

Note: The two special fittings, the "aerator" and the "deaerator" are the basis for the self-venting features of "Sovent."

The "aerator" does three things:

1. It limits the velocity of both liquid and air in the stack.

2. It prevents the cross section of the stack from filling with a plug of water.

3. It efficiently mixes the waste flowing in the branches with the air in the stack.

The "Sovent Aerator Fitting" mixes waste and air so effectively that no plug of water can form across the stack diameter and disturb fixture trap seals.

At a floor level where no "aerator" fitting is needed a double in-line offset is used.

Note: See drawing below.

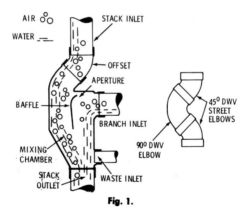

Fig. 1.

COPPER SOVENT
THE SOVENT AERATOR FITTING

Consists of an offset at the upper stack inlet connection, a mixing chamber, one or more branch inlets, one or more waste inlets for connection of smaller waste branches, a baffle in the center of the chamber with an aperture between it and the top of the fitting. and the stack outlet at the bottom of the fitting.

The "aerator" fitting provides a chamber where the flow of soil and waste from horizontal branches can unite smoothly with the air and liquid already flowing in the stack. The "aerator" fitting prevents pressure and vacuum fluctuations that could blow or siphon fixtures trap seals.

Note: No "aerator" fitting is needed at a floor level where no soil branch enters and only a 2-inch (51 mm) waste branch enters a 4-inch (102 mm) stack.

A double in line offset is used in place of the "aerator" fitting. This offset reduces the fall rate in the stack between floor intervals in a manner similar to the "aerator" fitting.

Note See Figure (1).

The "Sovent" deaerator fitting relieves the pressure build-up at the bottom of the stack, venting that pressure into a relief line that connects into the top of the building drain. The "deaerator" pressure relief line is tied to the building drain or, at an offset to the lower stack.

Note: See Figure (3), also related drawing Figure (2).

2. The SOVENT deaerator fitting relieves the pressure build-up at the bottom of the stack, venting that pressure into a relief line that connects into the top of the building drain. The deaerator pressure relief line is tied to the building drain or, at an offset to the lower stack.

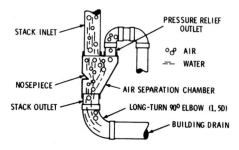
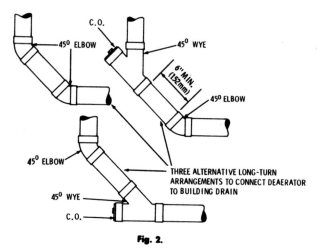

Fig. 2.

THE "SOVENT DEAERATOR FITTING"

Consists of an air separation chamber having an internal nosepiece, a stack inlet, a pressure relief outlet at the top, and a stack outlet at the bottom.

Copper Sovent

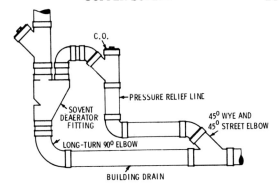

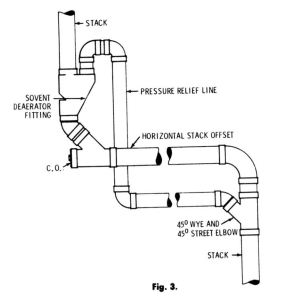

Fig. 3.

The "deaerator" fitting at the bottom of the stack functions in combination with the "aerator" fitting above to make the single stack self-venting.

The "deaerator" is designed to overcome the tendency that would otherwise occur for the falling waste to build up excessive back pressure at the bottom of the stack when the flow is decelerated by the bend into the horizontal drain. The "deaerator" provides a method of separating air from system flow and equalizes pressure buildups.

Tests show that the configuration of the "deaerator" fitting causes part of the air falling with the liquid and solid in the stack to be ejected through the pressure relief line to the top of the building drain while the balance goes into the drain with the soil and waste.

At the "deaerator" outlet, the stack is connected into the horizontal drain through a long-turn fitting arrangement. Downstream at least 4 feet (122 cm) from this point the "pressure relief line" from the top of the "deacrator" fitting is connected into the top of the building drain.

A "deaerator" fitting with its pressure relief line connection, is installed in this way at the base of every "Sovent" stack and also at every offset (vertical-horizontal-vertical) in a stack.

In the latter case, the "pressure relief line" is tied into the stack immediately below the horizontal portion.

Note: Refer to Figure (4).

4. Copper SOVENT drainage stack design features.

COPPER SOVENT

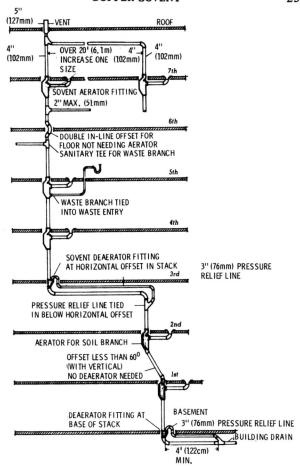

Fig. 4.

THE STACK

The stack must be carried full size through the roof.

Two stacks can be tied together at the top above the highest fixture; one stack extending thru the roof.

If the distance between the two stacks is 20 feet (6.1 m) or less, the horizontal tie-line can be the same diameter as the stack that terminates below the roof level.

If the distance is greater than 20 feet (6.1 m) the tie-line must be one size larger than the terminated stack.

The common stack extending through the roof must be one pipe size larger than the size of the larger stack below their tie-line.

The size of the stack is determined by the number of fixture units connected, as with traditional sanitary systems.

Existing "sovent" installations include 4-inches (102 mm) stacks serving up to 400 fixture units and 30 stories in height.

"Sovents" (Cost Saving Potential) can be seen by considering the 12-story stack illustration serving an apartment grouping.

The material saving is shown graphically in the schematic riser diagrams for "two-pipe" and "sovent" systems.

Note: Refer to Figure (5).

5. SOVENT's cost-saving potential can be seen by considering a 12-story stack to serve this apartment grouping. The material saving is shown graphically in the schematic riser diagrams for two-pipe and SOVENT systems.

COPPER SOVENT

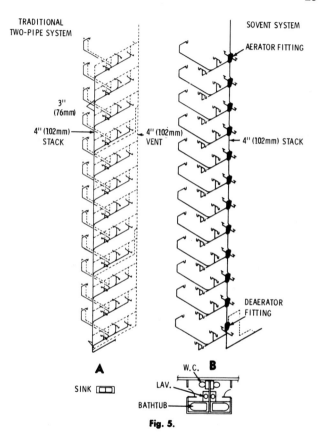

Fig. 5.

6. Following the SOVENT design rules results in light-weight efficient branches.

Copper Sovent

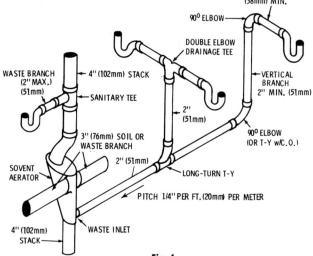

Fig. 6.

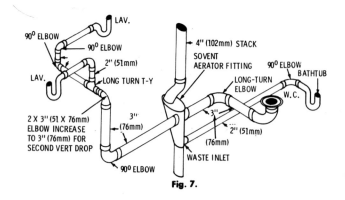

Fig. 7.

COPPER SOVENT

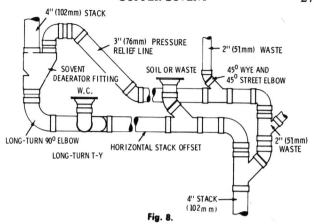

Fig. 8.

8. Soil and waste branches may be connected into a horizontal stack offset. Waste branches may be connected into the pressure relief line.

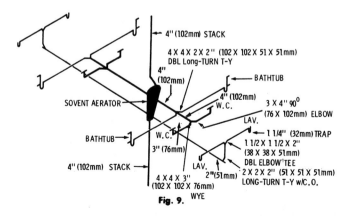

Fig. 9.

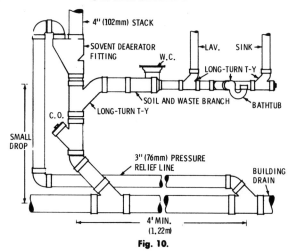

Fig. 10.

10. Soil and waste branches may be connected immediately below a deaerator fitting at the bottom of the stack.

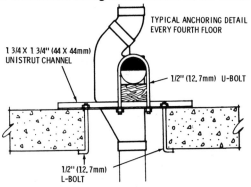

Fig. 11.

COPPER SOLVENT

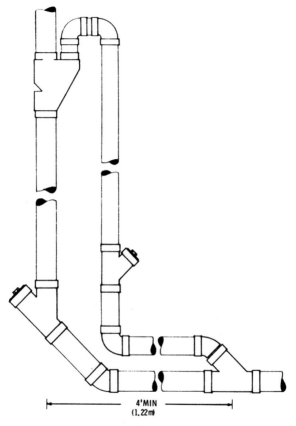

Fig. 12.

12. The deaerator fitting may be located above the floor level of the building drain.

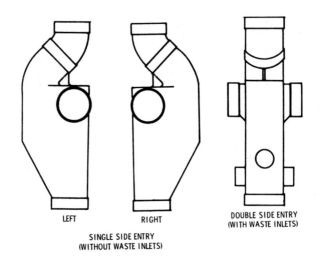

Fig. 13.

Fig. 14. Installing a copper solvent system.

Copper Drainage Fittings and Specifications

I wish to thank Nibco, Inc., Elkhart, Indiana, for making this section on "Copper Drainage Fittings" possible.

DRAINAGE FITTINGS

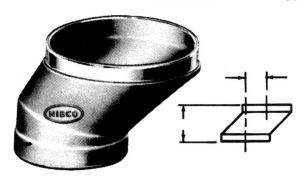

OFFSET CLOSET FITTING—FTGXC

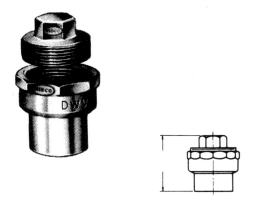

DWV—FTG. X CLEANOUT W/PLUG

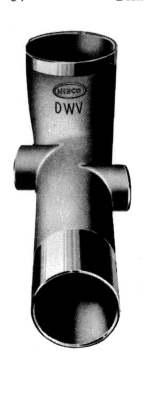
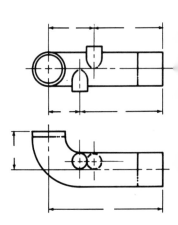

DWV—CLOSET BEND—R-L INLET

Drainage Fittings

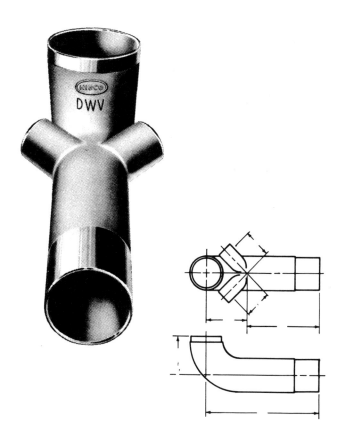

DWV CLOSET BEND — RIGHT & LEFT INLET

DRAINAGE FITTINGS

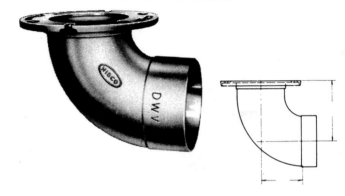

DWV—90° CLOSET ELL

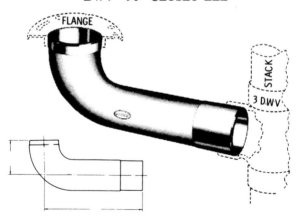

DWV CLOSET BEND FTG. X FTG.

DRAINAGE FITTINGS

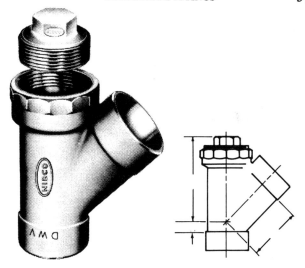

DWV — 45° Y — WITH C.O. PLUG

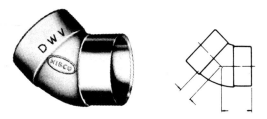

DWV — 45° — FTG. X COPPER ELL

DRAINAGE FITTINGS

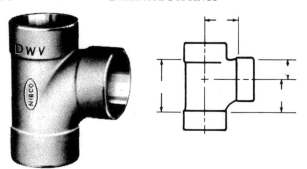

DWV — TEE CXCXC

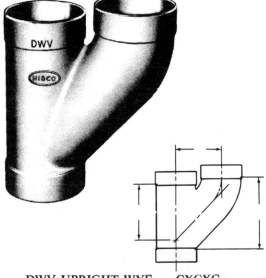

DWV UPRIGHT WYE — CXCXC

DRAINAGE FITTINGS

39

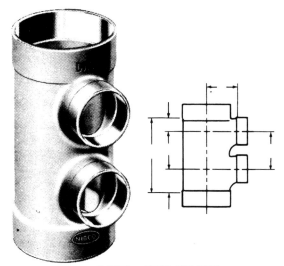

DWV — TEE CXCXC
DWV STACK FITTING — W/2-SIDE INLETS

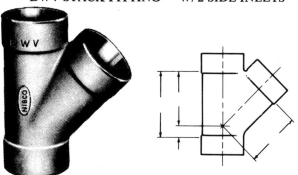

DWV - 45° - Y - BRANCH — CXCXC

DRAINAGE FITTINGS

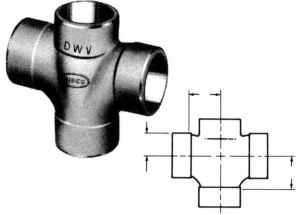

DWV VENT CROSS CXCXCXC

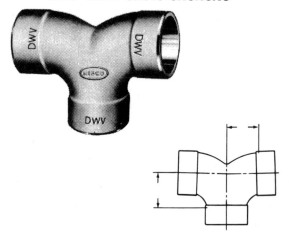

DWV TWIN ELL CXCXC

DRAINAGE FITTINGS

DWV TEST TEE — CXC — W/PLUG

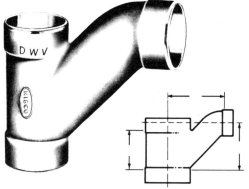

DWV LONG TURN T-Y — CXCXC

VENTS AND VENTING ILLUSTRATED

"Continuous Vent" (also back vent) is a vertical vent that is a continuation of the drain to which it connects.

"Main Vent" same as "Vent Stack" is the principal artery of the venting system to which vent branches may be connected.

"Branch Vent" is a vent pipe connecting one or more individual vents with a vent stack, or stack vent.

"Wet Vent" is a waste pipe that also serves as an air circulating pipe or vent.

"Circuit Vent" is a branch vent that serves two or more traps and extends from in front of the last fixture connection of a horizontal branch _____ to the "Vent Stack".

"Individual Vent" is a vent pipe installed to vent a fixture trap. It may connect with another vent pipe 42 inches (1.07 m) or higher above fixture served, or terminate thru the roof individually.

"Dual Vent" also called "Common, or Unit Vent" is a vent connecting at the junction of two fixture drains and acting as a vent for both fixtures.

A "Relief or Yoke Vent" is a vent the main function being to provide circulation of air between drainage and vent system.

A "Local Vent" is a ventilating pipe on the fixture inlet side of the trap; this vent permits vapor or foul air to be removed from a fixture or room. This removal of foul air or offensive odors from toilet rooms is accomplished now by bathroom ventilation, fans and ducts.

A "Dry Vent" conducts air and vapor only to the open air.

VENTS AND VENTING
EXAMPLE IN VENTING

Note: Fresh-air system is installed for ice machine, steam table, and vegetable sink.

Note: Garbage disposal is placed in sanitary system; is not permitted on fresh air system.

Note: Sewage ejector vent, and fresh air auxiliary vent extend thru roof independently.

Note: Walls etc. eliminated for simplicity.

Note:
C.O. = Cleanout
W.C. = Water Closet
V.T.R. = Vent Thru Roof

VENT & DRAINAGE SYSTEM "RESIDENCE"

Note: If garbage disposal is installed on sink, horizontal branch waste line should be 3 inches (76 mm) and continue up to sink opening. Check local code.

Vents and Venting

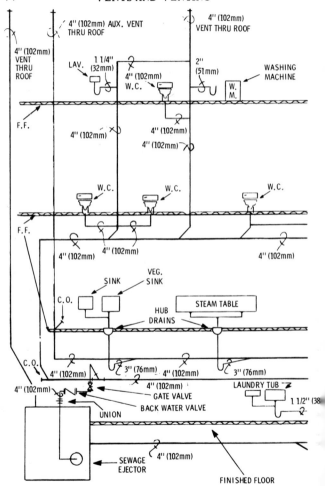

Vents and Venting

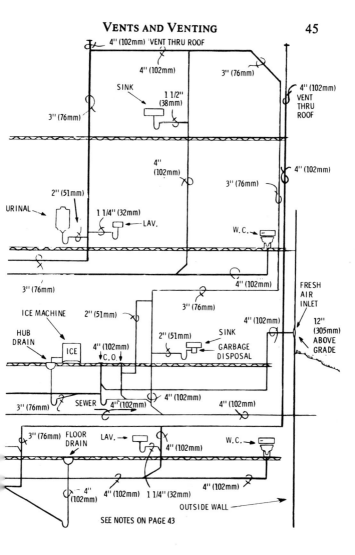

SEE NOTES ON PAGE 43

VENTS AND VENTING

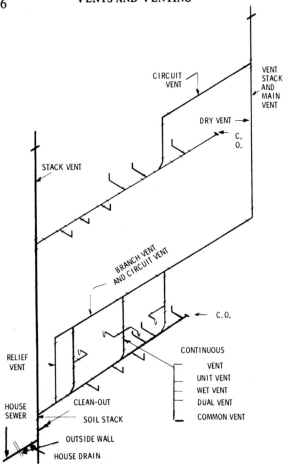

Water Closets & Lavatories Vented

VENTS AND VENTING

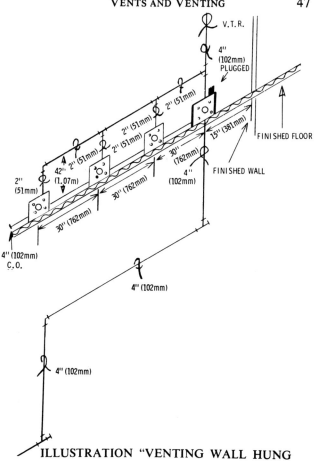

ILLUSTRATION "VENTING WALL HUNG WATER CLOSETS"
"Example of Chair Carrier Installation"

48 VENTS AND VENTING

V.T.R.
STACK VENT

LOOP VENT

RELIEF VENT
UNIT VENT
WET-VENT
CONTINUOUS VENT

RELIEF VENT

SOIL STACK

CLEANOUT

Water Closets and Lavatories Vented

VENTS AND VENTING

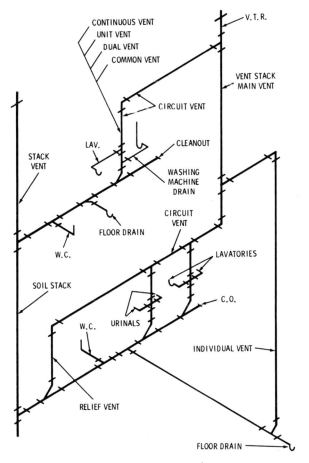

Example in Venting

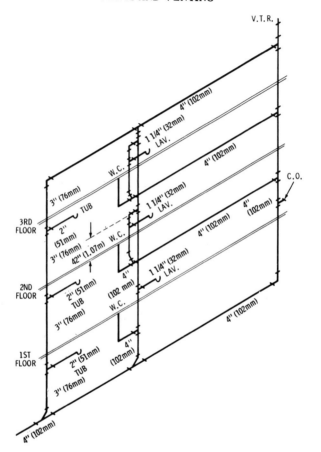

VENTING ILLUSTRATION
Tubs, Water Closets and Lavatories Vented

VENTS AND VENTING 51

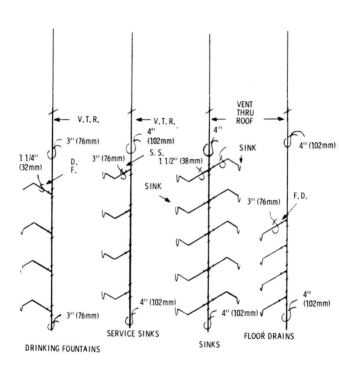

VENTING ILLUSTRATIONS
Combination Waste and Vent Stacks

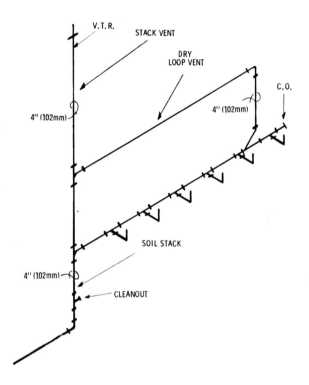

Water Closets Vented

Vents and Venting

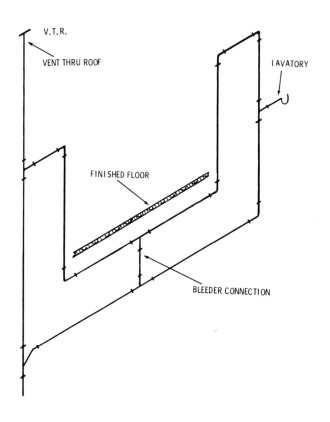

Looped Vent

VENTS AND VENTING

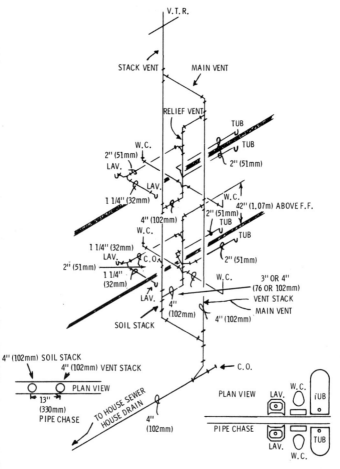

WATER CLOSETS, TUBS, LAVATORIES VENTED

INTRODUCTION TO "CAST-IRON SOIL PIPE INSTITUTE"

The two sections in this book entitled "Cast-Iron Pipe Fittings" and "Cast-Iron No-Hub Pipe Fittings" are to me the most helpful and beneficial sections of my entire Handbook. These sections could not have been made possible without the permission of the "Cast-Iron Soil Pipe Institute."

They are to be commended for their fine contribution to the plumbing industry.

Their efforts and standardization of cast iron fittings has made our job as plumbers a great deal easier.

The drawings and specifications contained herein have been taken from the "Cast-Iron Soil Pipe and Fitting Handbook."

Members of the "Cast-Iron Soil Pipe Institute" adopted an insignia ₵ for use as a symbol of quality and to provide a simple method of designating the standard desired. The standardization program carried on by the C.I.S.P.I. has raised the quality of cast-iron soil pipe and fittings, and today "Service Wt. Pipe and Fittings" bearing the ₵ insignia are a better product than "Extra Heavy Pipe and Fittings" were about twenty year ago. At the present time the CISPI consists of 12 companies, operating 16 plants in 8 states.

Members of the Cast-Iron Soil Pipe Institute
Manufacturers of C. I. Soil Pipe & Fittings

The American Brass and Iron Foundry ....Oakland, Cal.
415-632-3467

American Foundry	Los Angeles, Cal.
	213-728-1788
Anaheim Foundry Company	Anaheim, Cal.
	714-870-9000
The Central Foundry Company	Tuscaloosa, Ala.
	205-553-6810
	Quakertown, Pa.
	215-536-2585
Charlotte Pipe and Foundry Company	Charlotte, N.C.
	704-372-5030
The Eastern Foundry Company	Boyertown, Pa.
	215-367-2153
Griffin Pipe Company	Lynchburg, Va.
	804-845-8021
Jones Mfg. Co.	Birmingham, Ala.
	205-956-5511
Southeastern Specialty and Mfg. Co.	Anniston, Ala.
	205-237-6641
Tyler Pipe Industries	Tyler, Texas
	214-882-5511
	Macungie, Pa.
	215-967-5141
United States Pipe and Foundry Co.	Birmingham, Ala.
	205-254-7000
	Chattanooga, Tenn.
	615-265-4611
	Anniston, Ala.
	205-831-3660
Universal Cast-Iron Mfg. Co.	South Gate, Cal.
	213-569-8151

CAST-IRON SOIL PIPE FITTINGS AND SPECIFICATIONS

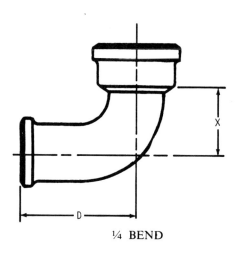

¼ BEND

Size	"D"	"X"
2"	6"	3¼"
3"	7"	4"
4"	8"	4½"
5"	8½"	5"
6"	9"	5½"
8"	11½"	6⅝"
10"	12½"	7⅝"

CAST-IRON SOIL PIPE FITTINGS AND SPECIFICATIONS

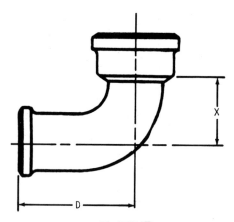

¼ BEND
"METRIC"

Size	"D"	"X"
51 mm	152 mm	83 mm
76 mm	178 mm	102 mm
102 mm	203 mm	114 mm
127 mm	216 mm	127 mm
152 mm	229 mm	140 mm
203 mm	292 mm	168 mm
254 mm	318 mm	194 mm

Cast-Iron Pipe Fittings

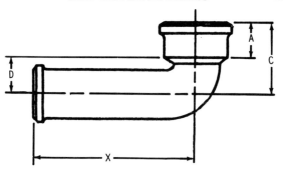

LONG LOW HUB ¼ BENDS

Size	"D"	"A"	"X"	"C"
4" x 12"	12"	3"	2¾"	5¾"
4" x 14"	14"	3"	2¾"	5¾"
4" x 16"	16"	3"	2¾"	5¾"
4" x 18"	18"	3"	2¾"	5¾"

Note:

Where space is limited in reference to water closets 1¾" is gained in measurement "C".

CAST-IRON PIPE FITTINGS

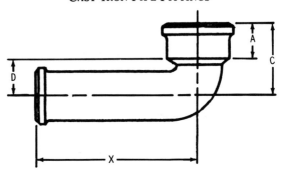

LONG LOW HUB ¼ BENDS
"METRIC"

Size	"D"	"A"	"X"	"C"
102 x 305 mm	305 mm	76 mm	70 mm	146 mm
102 x 356 mm	356 mm	76 mm	70 mm	146 mm
102 x 406 mm	406 mm	76 mm	70 mm	146 mm
102 x 457 mm	457 mm	76 mm	70 mm	146 mm

Note: Where space is limited in reference to water closets, 44 mm is gained in measurement "C".

Cast-Iron Pipe Fittings

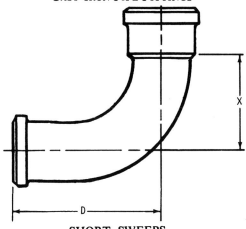

SHORT SWEEPS

Size	"D"	"X"
2"	8 "	5¼ "
3"	9 "	6 "
4"	10 "	6½ "
5"	10½"	7 "
6"	11 "	7½ "
8"	13½"	8⅝ "
10"	14½"	9⅝ "

LONG SWEEPS

Size	"D"	"X"
2"	11 "	8¼ "
3"	12 "	9 "
4"	13 "	9½ "
5"	13½"	10 "
6"	14 "	10½ "
8"	16½"	11⅝ "
10"	17½"	12⅝ "

REDUCING LONG SWEEPS

Size	"D"	"X"
3" x 2"	9"	6 "
4" x 3"	10"	6½"

Cast-Iron Pipe Fittings

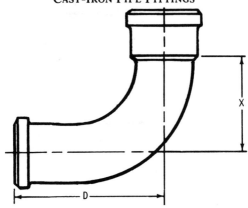

SHORT SWEEPS "METRIC"

Size	"D"	"X"
51 mm	203 mm	133 mm
76 mm	229 mm	152 mm
102 mm	254 mm	165 mm
127 mm	267 mm	178 mm
152 mm	279 mm	191 mm
203 mm	343 mm	219 mm
254 mm	368 mm	244 mm

LONG SWEEPS "METRIC"

Size	"D"	"X"
51 mm	279 mm	210 mm
76 mm	305 mm	229 mm
102 mm	330 mm	241 mm
127 mm	343 mm	254 mm
152 mm	356 mm	267 mm
203 mm	419 mm	295 mm
254 mm	445 mm	321 mm

REDUCING LONG SWEEPS "METRIC"

Size	"D"	"X"
76 x 51 mm	229 mm	152 mm
102 x 76 mm	254 mm	165 mm

Cast-Iron Pipe Fittings

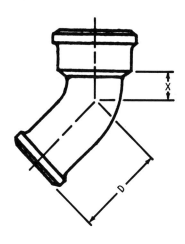

TABLE 1—⅛ Bends

Size	"D"	"X"
2"	4- 1/4 "	1- 1/2 "
3"	4-15/16"	1-15/16"
4"	5-11/16"	2- 3/16"
5"	5- 7/8 "	2- 3/8 "
6"	6- 1/16"	2- 9/16"
8"	8 "	3- 1/8 "
10"	8- 3/8 "	3- 1/2 "

TABLE 2—1/16 BENDS

Size	"D"	"X"
2"	3-5/8"	7/8"
3"	4-3/16"	1-3/16"
4"	4-13/16"	1-5/16"
5"	4-7/8"	1-3/8"
6"	5	1-1/2"
8"	6-11/16"	1-13/16"
10"	6-7/8"	2

TABLE 3—1/6 BENDS

Size	"D"	"X"
2"	4-3/4"	2
3"	5-1/2"	2-1/2"
4"	6-5/16"	2-13/16"
5"	6-5/8"	3-1/8"
6"	6-7/8"	3-3/8"
8"	9	4-1/8"
10"	9-9/16"	4-11/16"

Cast-Iron Pipe Fittings

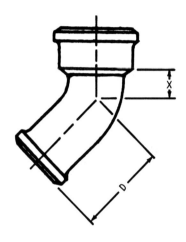

TABLE 1 — ⅛ BENDS
"METRIC"

Size	"D"	"X"
51 mm	108 mm	38 mm
76 mm	125 mm	49 mm
102 mm	144 mm	56 mm
127 mm	149 mm	60 mm
152 mm	154 mm	65 mm
203 mm	203 mm	79 mm
254 mm	213 mm	89 mm

TABLE 2 — 1/16 BENDS
"METRIC"

Size	"D"	"X"
51 mm	92 mm	22 mm
76 mm	106 mm	30 mm
102 mm	122 mm	33 mm
127 mm	124 mm	35 mm
152 mm	127 mm	38 mm
203 mm	170 mm	46 mm
254 mm	175 mm	51 mm

TABLE 3 — 1/6 BENDS
"METRIC"

Size	"D"	"X"
51 mm	121 mm	51 mm
76 mm	140 mm	64 mm
102 mm	160 mm	71 mm
127 mm	168 mm	79 mm
152 mm	175 mm	86 mm
203 mm	229 mm	105 mm
254 mm	243 mm	119 mm

Cast-Iron Pipe Fittings

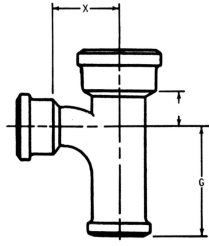

SINGLE & DOUBLE SANITARY T-BRANCHES

Size	"X"	"G"	"A"
2"	2¾"	6¼"	1¾"
3"	4"	7½"	2½"
4"	4½"	8"	3"
5"	5"	8½"	3½"
6"	5½"	9"	4"
3" x 2"	4"	7"	2"
4" x 2"	4½"	7"	2"
4" x 3"	4½"	7½"	2½"
5" x 2"	5"	7"	2"
5" x 3"	5"	7½"	2½"
5" x 4"	5"	8"	3"
6" x 2"	5½"	7"	2"
6" x 3"	5½"	7½"	2½"
6" x 4"	5½"	8"	3"
6" x 5"	5½"	8½"	3½"

CAST-IRON PIPE FITTINGS

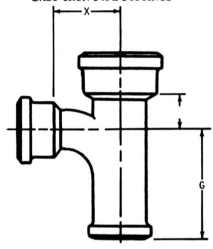

SINGLE & DOUBLE SANITARY T-BRANCHES
"METRIC"

Size	"X"	"G"	"A"
51 mm	70 mm	159 mm	44 mm
76 mm	102 mm	191 mm	64 mm
102 mm	114 mm	203 mm	76 mm
127 mm	127 mm	216 mm	89 mm
152 mm	140 mm	229 mm	102 mm
76 × 51 mm	102 mm	178 mm	51 mm
102 × 51 mm	114 mm	178 mm	51 mm
102 × 76 mm	114 mm	191 mm	64 mm
127 × 51 mm	127 mm	178 mm	51 mm
127 × 76 mm	127 mm	191 mm	64 mm
127 × 102 mm	127 mm	203 mm	76 mm
152 × 51 mm	140 mm	178 mm	51 mm
152 × 76 mm	140 mm	191 mm	64 mm
152 × 102 mm	140 mm	203 mm	76 mm
152 × 127 mm	140 mm	216 mm	89 mm

CAST-IRON PIPE FITTINGS

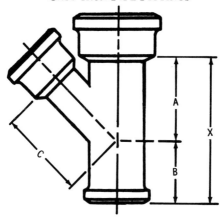

Y-BRANCHES, SINGLE & DOUBLE

Size	"A"	"B"	"C"	"X"
2"	4 "	4 "	4 "	8 "
3"	5-1/2 "	5 "	5-1/2 "	10-1/2 "
4"	6-3/4 "	5-1/4 "	6-3/4 "	12 "
5"	8 "	5-1/2 "	8 "	13-1/2 "
6"	9-1/4 "	5-3/4 "	9-1/4 "	15 "
8"	11-13/16 "	7-11/16"	11-13/16"	19-1/2 "
10"	14-1/2 "	8 "	14-1/2 "	22-1/2 "

Size	"A"	"B"	"C"	"X"
3" x 2"	4-13/16"	4-3/16"	5 "	9 "
4" x 2"	5-3/8 "	3-5/8 "	5¾"	9 "
4" x 3"	6-1/16"	4-7/16"	6¼"	10½"
5" x 2"	5-7/8 "	3-1/8 "	6½"	9 "
5" x 3"	6-5/8 "	3-7/8 "	7 "	10½"
5" x 4"	7-5/16"	4-11/16"	7½"	12 "
6" x 2"	6-7/16"	2-9/16"	7¼"	9 "
6" x 3"	7-1/8 "	3-3/8 "	7¾"	10½"
6" x 4"	7-13/16"	4-3/16"	8¼"	12 "
6" x 5"	8-9/16"	4-15/16"	8¾"	13½"

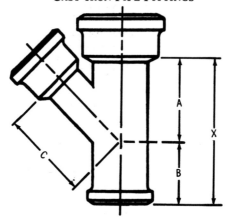

Y-BRANCHES, SINGLE & DOUBLE
"METRIC"

Size	"A"	"B"	"C"	"X"
51 mm	102 mm	102 mm	102 mm	203 mm
76 mm	140 mm	127 mm	140 mm	267 mm
102 mm	171 mm	133 mm	171 mm	305 mm
127 mm	203 mm	140 mm	203 mm	343 mm
152 mm	235 mm	146 mm	235 mm	381 mm
203 mm	300 mm	195 mm	300 mm	495 mm
254 mm	368 mm	203 mm	368 mm	572 mm
Size	"A"	"B"	"C"	"X"
76 × 51 mm	122 mm	106 mm	127 mm	229 mm
102 × 51 mm	137 mm	92 mm	146 mm	229 mm
102 × 76 mm	154 mm	113 mm	159 mm	267 mm
127 × 51 mm	149 mm	79 mm	165 mm	229 mm
127 × 76 mm	168 mm	98 mm	178 mm	267 mm
127 × 102 mm	186 mm	119 mm	191 mm	305 mm
152 × 51 mm	164 mm	65 mm	184 mm	229 mm
152 × 76 mm	181 mm	86 mm	197 mm	267 mm
152 × 102 mm	198 mm	106 mm	210 mm	305 mm
152 × 127 mm	217 mm	125 mm	222 mm	343 mm

Cast-Iron Pipe Fittings

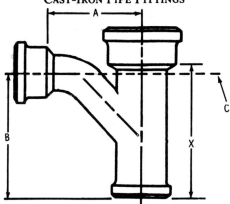

SINGLE & DOUBLE COMBINATION Y AND ⅛ BEND

Size	"A"	"B"	"C"	"X"
2"	4⅞"	7- 3/8"	5/8"	8"
3"	7"	10- 1/16"	7/16"	10½"
4"	9"	12"	0"	12"
5"	11"	14- 1/8"		13½"
6"	12⅞"	16- 1/16"		15"
8"	17"	21- 9/16"		19½"
3" x 2"	5¾"	8- 3/16"	13/16"	9"
4" x 2"	6¼"	8- 3/16"	13/16"	9"
4" x 3"	7½"	10- 1/16"	7/16"	10½"
5" x 2"	6¾"	8- 3/8"	5/8"	9"
5" x 3"	8"	10- 1/16"	7/16"	10½"
5" x 4"	9½"	12"	0"	12"
6" x 2"	7¼"	8- 3/16"	13/16"	9"
6" x 3"	8½"	10- 1/16"	7/16"	10½"
6" x 4"	10"	12"	0"	12"
6" x 5"	11½"	14- 3/16"		13½"

Cast-Iron Pipe Fittings

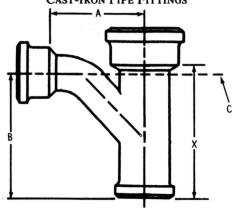

SINGLE & DOUBLE COMBINATION Y AND ⅛ BEND
"METRIC"

Size	"A"	"B"	"C"	"X"
51 mm	124 mm	187 mm	16 mm	203 mm
76 mm	178 mm	256 mm	11 mm	267 mm
102 mm	229 mm	305 mm	0 mm	305 mm
127 mm	279 mm	359 mm		343 mm
152 mm	327 mm	408 mm		381 mm
203 mm	432 mm	548 mm		495 mm
76 × 51 mm	146 mm	208 mm	21 mm	229 mm
102 × 51 mm	159 mm	208 mm	21 mm	229 mm
102 × 76 mm	191 mm	256 mm	11 mm	267 mm
127 × 51 mm	171 mm	213 mm	16 mm	229 mm
127 × 76 mm	203 mm	256 mm	11 mm	267 mm
127 × 102 mm	241 mm	305 mm	0 mm	305 mm
152 × 51 mm	184 mm	208 mm	21 mm	229 mm
152 × 76 mm	216 mm	256 mm	11 mm	267 mm
152 × 102 mm	254 mm	305 mm	0 mm	305 mm
152 × 127 mm	292 mm	360 mm		343 mm

Cast-Iron Pipe Fittings

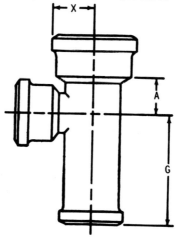

SINGLE & DOUBLE "T"-BRANCHES

Size	"X"	"G"	"A"
2"	1¾"	6¼"	1¾"
3"	2½"	7½"	2½"
4"	3 "	8 "	3 "
5"	3½"	8½"	3½"
6"	4 "	9 "	4 "
3" x 2"	2½"	7 "	2 "
4" x 2"	3 "	7 "	2 "
4" x 3"	3 "	7½"	2½"
5" x 2"	3½"	7 "	2 "
5" x 3"	3½"	7½"	2½"
5" x 4"	3½"	8 "	3 "
6" x 2"	4 "	7 "	2 "
6" x 3"	4 "	7½"	2½"
6" x 4"	4 "	8 "	3 "
6" x 5"	4 "	8½"	3½"

CAST-IRON PIPE FITTINGS

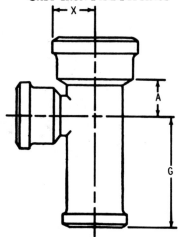

SINGLE & DOUBLE "T"—BRANCHES
"METRIC"

Size	"X"	"G"	"A"
51 mm	44 mm	159 mm	44 mm
76 mm	64 mm	191 mm	64 mm
102 mm	76 mm	203 mm	76 mm
127 mm	89 mm	216 mm	89 mm
152 mm	102 mm	229 mm	102 mm
76 × 51 mm	64 mm	178 mm	51 mm
102 × 51 mm	76 mm	178 mm	51 mm
102 × 76 mm	76 mm	191 mm	64 mm
127 × 51 mm	89 mm	178 mm	51 mm
127 × 76 mm	89 mm	191 mm	64 mm
127 × 102 mm	89 mm	203 mm	76 mm
152 × 51 mm	102 mm	178 mm	51 mm
152 × 76 mm	102 mm	191 mm	64 mm
152 × 102 mm	102 mm	203 mm	76 mm
152 × 127 mm	102 mm	216 mm	89 mm

Cast-Iron Pipe Fittings

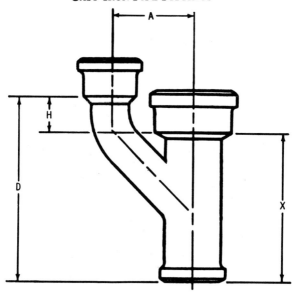

UPRIGHT "Y" BRANCHES

Size	"A"	"H"	"X"	"D"
2"	4½"	2 "	8 "	10 "
3"	5½"	1-15/16"	10½"	12-7/16"
4"	6½"	1-15/16"	12 "	13-15/16"
3" x 2"	5 "	1-15/16"	9 "	10-15/16"
4" x 2"	5½"	1-15/16"	9 "	10-15/16"
4" x 3"	6 "	1-15/16"	10½"	12-7/16"

Cast-Iron Pipe Fittings

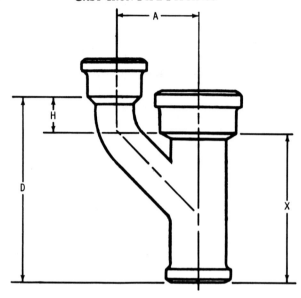

UPRIGHT "Y" BRANCHES
"METRIC"

Size	"A"	"H"	"X"	"D"
51 mm	114 mm	51 mm	203 mm	254 mm
76 mm	140 mm	49 mm	267 mm	316 mm
102 mm	165 mm	49 mm	305 mm	354 mm
76 × 51 mm	127 mm	49 mm	229 mm	278 mm
102 × 51 mm	140 mm	49 mm	229 mm	278 mm
102 × 76 mm	152 mm	49 mm	267 mm	316 mm

Cast-Iron Pipe Fittings

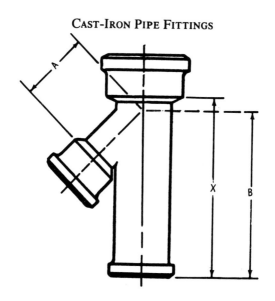

INVERTED "Y" BRANCHES

Size	"A"	"B"	"X"
2"	3⅜"	8¾"	9½"
3"	4⅝"	11¼"	12½"
4"	5⅞"	12½"	14 "
3" x 2"	4⅛"	10½"	11 "
4" x 2"	4⅞"	10-15/16"	11 "
4" x 3"	5⅜"	11¾"	12½"

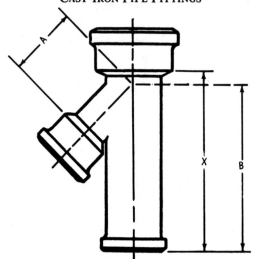

INVERTED "Y" BRANCHES
"METRIC"

Size	"A"	"B"	"X"
51 mm	86 mm	222 mm	241 mm
76 mm	117 mm	286 mm	318 mm
102 mm	149 mm	318 mm	356 mm
76 × 51 mm	105 mm	267 mm	279 mm
102 × 51 mm	124 mm	278 mm	279 mm
102 × 76 mm	137 mm	298 mm	318 mm

CAST-IRON PIPE FITTINGS

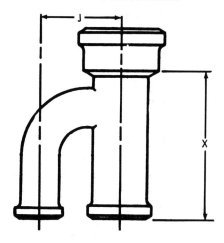

VENT-BRANCHES

Size	"J"	"X"
2"	4½"	8"
3"	5½"	10"
4"	6½"	11"
3" x 2"	5 "	9"
4" x 2"	5½"	9"
4" x 3"	6 "	10"

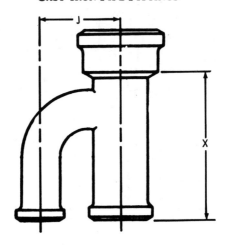

VENT BRANCHES
"METRIC"

Size	"J"	"X"
51 mm	114 mm	203 mm
76 mm	140 mm	254 mm
102 mm	165 mm	279 mm
76 × 51 mm	127 mm	229 mm
102 × 51 mm	140 mm	229 mm
102 × 76 mm	152 mm	254 mm

Cast-Iron Pipe Fittings

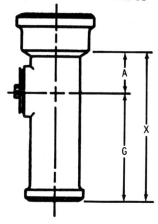

"CLEANOUT T" BRANCH

Size	"A"	"G"	"X"
2"	1¾"	6¼"	8"
3"	2½"	7½"	10"
4"	3 "	8 "	11"
5"	3½"	8½"	12"
6"	4 "	9 "	13"

Cast-Iron Pipe Fittings

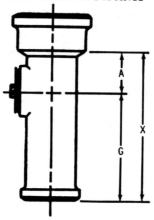

"CLEANOUT T" BRANCH
"METRIC"

Size	"A"	"G"	"X"
51 mm	44 mm	159 mm	203 mm
76 mm	64 mm	191 mm	254 mm
102 mm	76 mm	203 mm	279 mm
127 mm	89 mm	216 mm	305 mm
152 mm	102 mm	229 mm	330 mm

CAST-IRON PIPE FITTINGS 83

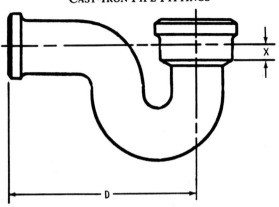

½ S OR P TRAPS

Size	"D"	"X"
2"	9½"	1½"
3"	12"	1¼"
4"	14"	1"
5"	15½"	½"
6"	17"	0"

Cast-Iron Pipe Fittings

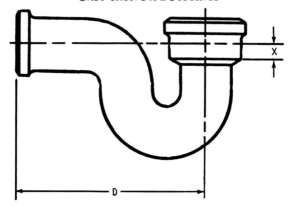

½ S OR P TRAPS
"METRIC"

Size	"D"	"X"
51 mm	241 mm	38 mm
76 mm	305 mm	32 mm
102 mm	356 mm	25 mm
127 mm	394 mm	13 mm
152 mm	432 mm	0 mm

Cast-Iron Pipe Fittings

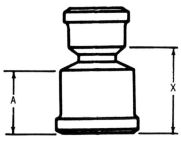

REDUCERS

Size	"A"	"X"
3" x 2"	3¼"	4¾"
4" x 2"	4 "	5 "
4" x 3"	4 "	5 "
5" x 2"	4 "	5 "
5" x 3"	4 "	5 "
5" x 4"	4 "	5 "
6" x 2"	4 "	5 "
6" x 3"	4 "	5 "
6" x 4"	4 "	5 "
6" x 5"	4 "	5 "

REDUCERS
"METRIC"

Size	"A"	"X"
76 × 51 mm	83 mm	121 mm
102 × 51 mm	102 mm	127 mm
102 × 76 mm	102 mm	127 mm
127 × 51 mm	102 mm	127 mm
127 × 76 mm	102 mm	127 mm
127 × 102 mm	102 mm	127 mm
152 × 51 mm	102 mm	127 mm
152 × 76 mm	102 mm	127 mm
152 × 102 mm	102 mm	127 mm
152 × 127 mm	102 mm	127 mm

Cast-Iron Pipe Fittings

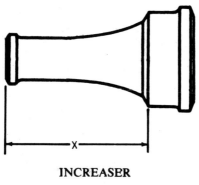

INCREASER

Size	"X"
2" x 3"	9"
2" x 4"	9"
2" x 5"	9"
2" x 6"	9"
3" x 4"	9"
3" x 5"	9"
3" x 6"	9"
4" x 5"	9"
4" x 6"	9"
4" x 8"	12"

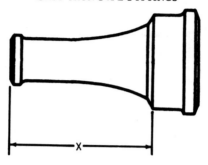

INCREASER
"METRIC"

Size	"X"
51 × 76 mm	229 mm
51 × 102 mm	229 mm
51 × 127 mm	229 mm
51 × 152 mm	229 mm
76 × 102 mm	229 mm
76 × 127 mm	229 mm
76 × 152 mm	229 mm
102 × 127 mm	229 mm
102 × 152 mm	229 mm
102 × 203 mm	305 mm

Cast-Iron Pipe Fittings

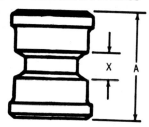

DOUBLE HUB FITTING

Size	"X"	"A"
2"	1"	6"
3"	1"	6½"
4"	1"	7"
5"	1"	7"
6"	1"	7"
8"	1¼"	8¼"
10"	1¼"	8¼"

CAST-IRON PIPE FITTINGS

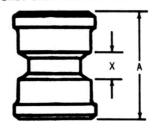

DOUBLE HUB FITTING
"METRIC"

Size	"X"	"A"
51 mm	25 mm	152 mm
76 mm	25 mm	165 mm
102 mm	25 mm	178 mm
127 mm	25 mm	178 mm
152 mm	25 mm	178 mm
203 mm	32 mm	210 mm
254 mm	32 mm	210 mm

Cast-Iron Pipe Fittings

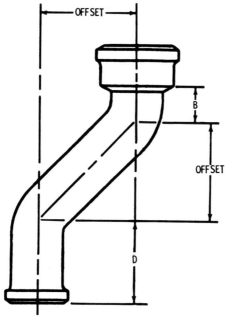

⅛ BEND OFFSET

Size	"Offset"	Hub	"D"	"B"
2"	(2"-4"-6"-8"-10")	2½"	4¼"	1"
3"	(2"-4"-6"-8"-10")	2¾"	5"	1½"
4"	(2"-4"-6"-8"-10")	3"	5¼"	1¾"
5"	(2"-4"-6"-8"-10")	3"	5-9/16"	1-15/16"
6"	(2")	3"	5⅝"	2"
6"	(4"-6"-8"-10")	3"	5-13/16"	2-3/16"

Note:
⅛ Bend Offset Fittings made in Pipe sizes 2" thru 6" (Offsets 2" thru 18")

Cast-Iron Pipe Fittings

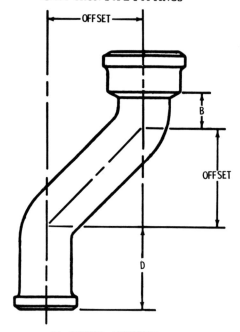

⅛ BEND OFFSET
"METRIC"

Size	Offset	Hub	"D"	"B"
51 mm	(51-102-152-203-254 mm)	64 mm	108 mm	25 mm
76 mm	(51-102-152-203-254 mm)	70 mm	127 mm	38 mm
102 mm	(51-102-152-203-254 mm)	76 mm	133 mm	44 mm
127 mm	(51-102-152-203-254 mm)	76 mm	141 mm	49 mm
152 mm	(51 mm)	76 mm	143 mm	51 mm
152 mm	(102-152-203-254 mm)	76 mm	149 mm	56 mm

Note:
 1/8 Bend offset fittings made in pipe sizes 51 thru 152 mm (Offsets 51 thru 457 mm).

Cast-Iron Pipe Fittings

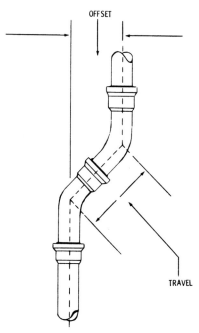

TABLE 1

MINIMUM OFFSETS USING ⅛th BEND FITTINGS

Size	Travel	Min. Offset
2"	5¾"	4- 3/32"
3"	6⅞"	4⅞ "
4"	7⅞"	5- 9/16"
5"	8¼"	5⅞ "
6"	8⅝"	6⅛ "
8"	11⅛"	7⅞ "
10"	11⅞"	8- 7/16"
12"	14⅜"	10¼ "

TABLE 2

MINIMUM OFFSETS USING 1/16th BEND FITTINGS

Size	Travel	Min. Offset
2"	4½"	1¾ "
3"	5⅜"	2-1/16"
4"	6⅛"	2-5/16"
5"	6¼"	2⅜ "
6"	6½"	2-7/16"
8"	8½"	3¼ "
10"	8⅞"	3⅜ "
12"	11 "	4¼ "

TABLE 3

MINIMUM OFFSETS USING 1/6th BEND FITTINGS

Size	Travel	Min. Offset
2"	6¾"	5⅞ "
3"	8 "	6-15/16"
4"	9⅛"	7-15/16"
5"	9¾"	8½ "
6"	10¼"	8⅞ "
8"	13⅛"	11⅜ "
10"	14¼"	12⅜ "
12"	17 "	14¾ "

Cast-Iron Pipe Fittings

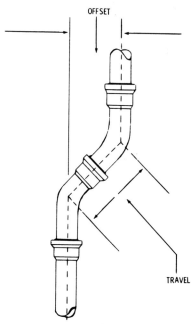

TABLE 1
MINIMUM OFFSETS USING ⅛th BEND FITTINGS
"METRIC"

Size	Travel	Min. Offset
51 mm	146 mm	104 mm
76 mm	175 mm	124 mm
102 mm	200 mm	141 mm
127 mm	210 mm	149 mm
152 mm	219 mm	156 mm
203 mm	283 mm	200 mm
254 mm	302 mm	214 mm
305 mm	365 mm	260 mm

TABLE 2
MINIMUM OFFSETS USING
1/16th BEND FITTINGS
"METRIC"

Size	Travel	Min. Offset
51 mm	114 mm	44 mm
76 mm	137 mm	52 mm
102 mm	156 mm	59 mm
127 mm	159 mm	60 mm
152 mm	165 mm	62 mm
203 mm	216 mm	83 mm
254 mm	225 mm	86 mm
305 mm	279 mm	108 mm

TABLE 3
MINIMUM OFFSETS USING
1/6th BEND FITTINGS
"METRIC"

Size	Travel	Min. Offset
51 mm	171 mm	149 mm
76 mm	203 mm	176 mm
102 mm	232 mm	202 mm
127 mm	248 mm	216 mm
152 mm	260 mm	225 mm
203 mm	333 mm	289 mm
254 mm	362 mm	314 mm
305 mm	432 mm	375 mm

CAST-IRON NO-HUB PIPE FITTINGS AND SPECIFICATIONS

The most outstanding advantages of ¢ NO-HUB joint:

Faster installation, more economical, space saving (3" (76 mm) size fits neatly in 2" × 4" (5 × 10 cm) framed wall), waste: absolutely none, quieter, vibrationless, lightweight, physically less taxing than conventional jointing, testing takes less time. Five floors can be tested at one time.

Recommended for inside and outside the house.

Limitations:

Additional brackets and supports may be needed for rigidity.

Introduction:

These suggestions are for use with the ¢ no-hub system utilizing a sleeve type coupling device consisting of an internally ribbed elastomeric sealing sleeve within a protective corrugated stainless steel shield band secured by two stainless steel bands with tightening devices also of stainless steel.

During installation assembly ¢ no-hub pipe and fittings must be inserted into sleeve and firmly seated against the center rib or shoulder of the gasket. In order to provide a sound joint with field cut lengths of pipe it is necessary to have ends cut smooth and as square as possible.

Snap type cutters may be used.

The stainless steel bands must be tightened alternately and firmly to not less than 48, nor more than 60 inch lbs. (67.79 N·m) of torque.

Installation:

Vertical Piping:

Secure vertical piping at sufficiently close intervals to keep the pipe in alignment and to support the weight of the pipe and its contents. Support stacks at their bases and at each floor.

Horizontal Piping Suspended:

Support ordinary horizontal piping and fittings at sufficiently close intervals to maintain alignment and prevent sagging or grade reversal.

Support each length of pipe by a hanger located as near coupling as possible but not more than 18" (46 cm) from the joint.

If piping is supported by non-rigid hangers more than 18 inches (46 cm) long, install sufficient sway bracing to prevent lateral movement, such as might be caused by seismic shock.

Hangers shall also be provided at each horizontal branch connection.

Horizontal Piping Underground:

₵ no-hub systems laid in trenches shall be continuously supported on undisturbed earth, or compacted fill of selected material, or on masonry blocks at each coupling.

To maintain proper alignment during backfilling, stabilize the pipe in proper position by partial back filling and cradling, or by the use of adequate metal stakes or braces fastened to the pipe.

Piping laid on grade should be adequately staked to prevent misalignment when the slab is poured.

Vertical sections and their connecting branches shall be adequately staked and fastened to driven pipe or reinforcing rod so as to remain stable while back fill is placed, or concrete is poured.

Remember: Spacer inside of neoprene gasket where fittings or pipe ends meet measures 3/32" (2.38 mm).

Note: Fitting measurements, laying lengths, may vary ⅛" (3.2 mm) plus or minus.

Note: 5 feet (152 cm) lengths of pipe may vary ¼" (6.4 mm) plus or minus.

10 feet (305 cm) lengths of pipe may vary ½" (13 mm) plus or minus.

Note: All 2", 3" and 4" (51, 76 and 102 mm) fittings, etc. lay the same length; this is to an advantage in that one fitting can be removed from a line and another inserted without cutting the pipe.

Cast-Iron No-Hub Fittings

NO-HUB Cast-Iron Pipe ready for joining. Note extreme simplicity of joint parts.

Sleeve coupling is placed on end of one pipe. Stainless steel shield and band clamps are placed on end of the other pipe.

CAST-IRON NO-HUB FITTINGS

Pipe ends are butted against integrally-molded shoulder inside of sleeve. Shield is slid into position and tightened to make a joint that is quickly assembled and permanently fastened.

Fittings are jointed and fastened in the same way. Making up joints in close quarters is easily done.

CAST-IRON NO-HUB FITTINGS 101

Extra support is needed with No-Hub. Note conventional hangers and supports in the ceiling to make the system more rigid.

Cast-Iron No-Hub Fittings

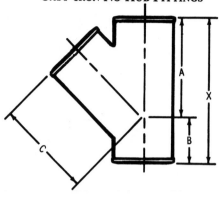

Y-BRANCHES, SINGLE & DOUBLE

Size	"A"	"B"	"C"	"X"
1½"	4 "	2 "	4 "	6 "
2"	4-5/8 "	2 "	4-5/8 "	6-5/8 "
3"	5-3/4 "	2-1/4 "	5-3/4 "	8 "
4"	7-1/16"	2-7/16 "	7-1/16"	9-1/2 "
5"	9-1/2 "	3-1/8 "	9-1/2 "	12-5/8 "
6"	10-3/4 "	3-5/16 "	10-3/4 "	14-1/16 "
3" x 2"	5-1/8 "	1-1/2 "	5-5/16"	6-5/8 "
4" x 2"	5-5/8 "	1 "	6 "	6-5/8 "
4" x 3"	6-5/16"	1-11/16"	6-1/2 "	8 "
5" x 2"	7-1/8 "	15/16"	7-1/2 "	8-1/16"
5" x 3"	8 "	1-11/16"	8 "	9-11/16"
5" x 4"	8-3/4 "	2-7/16 "	8-1/2 "	11-3/16"
6" x 2"	7-13/16"	1/2 "	8-1/4 "	8-5/16"
6" x 3"	8-1/2 "	1-1/4 "	8-3/4 "	9-3/4 "
6" x 4"	9-1/4 "	1-15/16 "	9-1/4 "	11-3/16"
6" x 5"	9-15/16"	2-9/16 "	10-1/4 "	12-1/2 "

Cast-Iron No-Hub Fittings

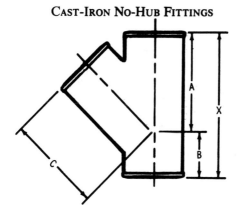

Y-BRANCHES, SINGLE & DOUBLE
"METRIC"

Size	"A"	"B"	"C"	"X"
38 mm	102 mm	51 mm	102 mm	152 mm
51 mm	117 mm	51 mm	117 mm	168 mm
76 mm	146 mm	57 mm	146 mm	203 mm
102 mm	179 mm	62 mm	179 mm	241 mm
127 mm	241 mm	79 mm	241 mm	321 mm
152 mm	273 mm	84 mm	273 mm	357 mm
76 × 51 mm	130 mm	38 mm	135 mm	168 mm
102 × 51 mm	143 mm	25 mm	152 mm	168 mm
102 × 76 mm	160 mm	43 mm	165 mm	203 mm
127 × 51 mm	181 mm	24 mm	191 mm	205 mm
127 × 76 mm	203 mm	43 mm	203 mm	246 mm
127 × 102 mm	222 mm	62 mm	216 mm	284 mm
152 × 51 mm	198 mm	13 mm	210 mm	211 mm
152 × 76 mm	216 mm	32 mm	222 mm	248 mm
152 × 102 mm	235 mm	49 mm	235 mm	284 mm
152 × 127 mm	252 mm	64 mm	260 mm	318 mm

Cast-Iron No-Hub Fittings

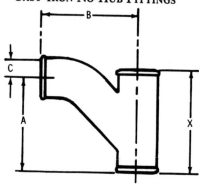

SINGLE & DOUBLE COMBINATION Y AND ⅛ BEND

Size	"A"	"B"	"C"	"X"
1½"	4- 3/4 "	5- 3/8 "	1- 1/4 "	6 "
2"	5- 3/8 "	6- 1/8 "	1- 1/4 "	6 5/8 "
3"	7- 5/16 "	8 "	11/16"	8 "
4"	9- 1/4 "	10 "	1/4 "	9- 1/2 "
5"	11- 3/4 "	12- 1/2 "	7/8 "	12- 5/8 "
6"	13- 5/8 "	14- 3/8 "	7/16 "	14- 1/16"
2" x 1½"	5 "	5- 5/8 "	1 "	6 "
3" x 2"	5- 1/2 "	6- 3/4 "	1- 1/8 "	6- 5/8 "
4" x 2"	5- 1/2 "	7- 1/4 "	1- 1/8 "	6- 5/8 "
4" x 3"	7- 1/4 "	8- 1/2 "	3/4 "	8 "
5" x 2"	5-15/16"	7- 3/4 "	2- 1/8 "	8- 1/16"
5" x 3"	7- 3/4 "	9 "	1-15/16"	9-11/16"
5" x 4"	9- 3/4 "	10- 1/2 "	1- 7/16 "	11- 3/16"
6" x 2"	6 "	8- 1/4 "	2- 5/16 "	8- 5/16"
6" x 3"	7-13/16"	9- 1/2 "	1-15/16"	9- 3/4 "
"6 x 4"	9- 3/4 "	11 "	1- 7/16 "	11- 3/16"
6" x 5"	11-11/16"	13 "	13/16 "	12- 1/2 "

Cast-Iron No-Hub Fittings

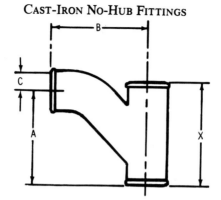

SINGLE & DOUBLE COMBINATION Y AND 1/8 BEND

"METRIC"

Size	"A"	"B"	"C"	"X"
38 mm	121 mm	137 mm	32 mm	152 mm
51 mm	137 mm	156 mm	32 mm	168 mm
76 mm	186 mm	203 mm	17 mm	203 mm
102 mm	235 mm	254 mm	6 mm	241 mm
127 mm	298 mm	318 mm	22 mm	321 mm
152 mm	346 mm	365 mm	11 mm	357 mm
51 × 38 mm	127 mm	143 mm	25 mm	152 mm
76 × 51 mm	140 mm	171 mm	29 mm	168 mm
102 × 51 mm	140 mm	184 mm	29 mm	168 mm
102 × 76 mm	184 mm	216 mm	19 mm	203 mm
127 × 51 mm	151 mm	197 mm	54 mm	205 mm
127 × 76 mm	197 mm	229 mm	49 mm	246 mm
127 × 102 mm	248 mm	267 mm	37 mm	284 mm
152 × 51 mm	152 mm	210 mm	59 mm	211 mm
152 × 76 mm	198 mm	241 mm	49 mm	248 mm
152 × 102 mm	248 mm	279 mm	37 mm	284 mm
152 × 127 mm	297 mm	330 mm	21 mm	318 mm

Cast-Iron No-Hub Fittings

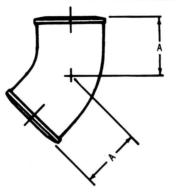

TABLE 1 — ⅛th BENDS

Size	"A"
1½"	2-5/8 "
2 "	2-3/4 "
3 "	3 "
4 "	3-1/8 "
5 "	3-7/8 "
6 "	4-1/16"

Note: Use above figure with following tables.

TABLE 2 — 1/16th BENDS

Size	"A"
2"	2- 1/8 "
3"	2- 1/4 "
4"	2- 5/16"
5"	2-15/16"
6"	3 "

TABLE 3 — 1/6th BENDS

Size	"A"
2"	3- 1/4 "
3"	3- 1/2 "
4"	3-13/16"

CAST-IRON NO-HUB FITTINGS

TABLE 4 — 1/5 BENDS

Size	"A"
2"	3-11/16"
3"	4-1/16"
4"	4-7/16"

TABLE 1 — 1/8th BENDS
"METRIC"

Size	"A"
38 mm	67 mm
51 mm	70 mm
76 mm	76 mm
102 mm	79 mm
127 mm	98 mm
152 mm	103 mm

Note: Use above figure with following tables.

TABLE 2 — 1/16TH BENDS
"METRIC"

Size	"A"
51 mm	54 mm
76 mm	57 mm
102 mm	59 mm
127 mm	75 mm
152 mm	76 mm

TABLE 3 — 1/6th BENDS
"METRIC"

Size	"A"
51 mm	83 mm
76 mm	89 mm
102 mm	97 mm

TABLE 4 — 1/5th BENDS
"METRIC"

Size	"A"
51 mm	94 mm
76 mm	103 mm
102 mm	113 mm

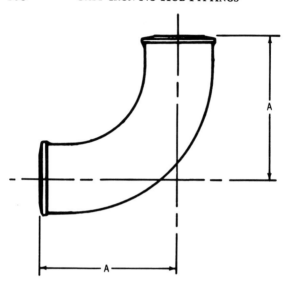

SHORT SWEEPS

Size	"A"
2"	6½"
3"	7 "
4"	7½"
5"	8½"
6"	9 "

Note: Use above figure with following tables.

REDUCING LONG SWEEPS

Size	"A"
3" x 2"	10 "
4" x 3"	10½"

CAST-IRON NO-HUB FITTINGS

LONG SWEEPS

Size	"A"
1½"	9¼"
2 "	9½"
3 "	10 "
4 "	10½"
5 "	11½"
6 "	12"

SHORT SWEEPS
"METRIC"

Size	"A"
51 mm	165 mm
76 mm	178 mm
102 mm	191 mm
127 mm	216 mm
152 mm	229 mm

Note: Use above figure with following tables.

LONG SWEEPS
"METRIC"

Size	"A"
38 mm	235 mm
51 mm	241 mm
76 mm	254 mm
102 mm	267 mm
127 mm	292 mm
152 mm	305 mm
203 mm	343 mm

REDUCING LONG SWEEPS
"METRIC"

Size	"A"
76 × 51 mm	254 mm
102 × 76 mm	267 mm

Cast-Iron No-Hub Fittings

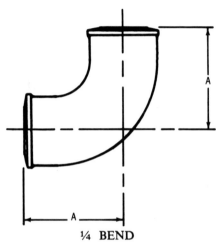

¼ BEND

Size	"A"
1½"	4¼"
2 "	4½"
3 "	5 "
4 "	5½"
5 "	6½"
6 "	7 "

¼ BEND "METRIC"

Size	"A"
38 mm	108 mm
51 mm	114 mm
76 mm	127 mm
102 mm	140 mm
127 mm	165 mm
152 mm	178 mm

Cast-Iron No-Hub Fittings

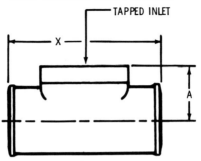

TEST TEES

Size	"X"	"A"
2"	6⅜"	2"
3"	7¾"	2-11/16"
4"	8⅞"	3"
5"	11½"	4½"
6"	12½"	5"

TEST TEE "METRIC"

Size	"X"	"A"
51 mm	162 mm	51 mm
76 mm	197 mm	68 mm
102 mm	225 mm	76 mm
127 mm	292 mm	114 mm
152 mm	318 mm	127 mm

Cast-Iron No-Hub Fittings

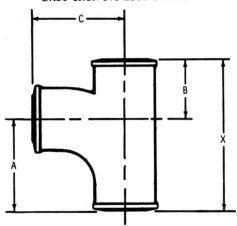

SINGLE & DOUBLE SANITARY T-BRANCHES

Size	"A"	"B"	"C"	"X"
1½"	4¼"	2¼"	4¼"	6½"
2 "	4½"	2⅜"	4½"	6⅞"
3 "	5 "	3 "	5 "	8 "
4 "	5½"	3⅝"	5½"	9⅛"
3" x 1½"	4¼"	2¼"	5 "	6½"
3" x 2 "	4½"	2⅜"	5 "	6⅞"
3" x 4 "	5½"	3½"	5 "	9 "
4" x 2 "	4½"	2⅜"	5½"	6⅞"
4" x 3 "	5 "	3 "	5½"	8 "
5" x 2 "	5 "	3½"	6½"	8½"

Cast-Iron No-Hub Fittings

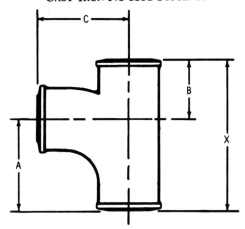

SINGLE & DOUBLE SANITARY T-BRANCHES
"METRIC"

Size	"A"	"B"	"C"	"X"
38 mm	108 mm	57 mm	108 mm	165 mm
51 mm	114 mm	60 mm	114 mm	175 mm
76 mm	127 mm	76 mm	127 mm	203 mm
102 mm	140 mm	92 mm	140 mm	232 mm
76 × 38 mm	108 mm	57 mm	127 mm	165 mm
76 × 51 mm	114 mm	60 mm	127 mm	175 mm
76 × 102 mm	140 mm	89 mm	127 mm	229 mm
102 × 51 mm	114 mm	60 mm	140 mm	175 mm
102 × 76 mm	127 mm	76 mm	140 mm	203 mm
127 × 51 mm	127 mm	89 mm	165 mm	216 mm

Cast-Iron No-Hub Fittings

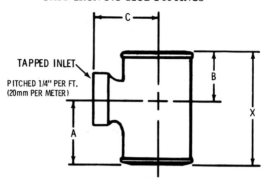

SINGLE & DOUBLE
SANITARY T-BRANCHES, TAPPED

Size	"A"	"B"	"C"	"X"
1½" x 1¼"	3¼"	2-7/16"	2- 9/16"	5-11/16"
1½" x 1½"	3¼"	2-7/16"	2- 9/16"	5-11/16"
2 " x 1¼"	3¼"	2-7/16"	2-13/16"	5-11/16"
2 " x 1½"	3¼"	2-7/16"	2-13/16"	5-11/16"
2 " x 2 "	3¾"	2-5/8 "	3- 1/16"	6- 3/8 "
3 " x 1¼"	3¼"	2-7/16"	3- 5/16"	5-11/16"
3 " x 1½"	3¼"	2-7/16"	3- 5/16"	5-11/16"
3 " x 2 "	3¾"	2-5/8 "	3- 9/16"	6- 3/8 "
4 " x 1¼"	3¼"	2-7/16"	3-13/16"	5-11/16"
4 " x 1½"	3¼"	2-7/16"	3-13/16"	5-11/16"
4 " x 2 "	3¾"	2-5/8 "	4- 1/16"	6- 3/8 "

Cast-Iron No-Hub Fittings

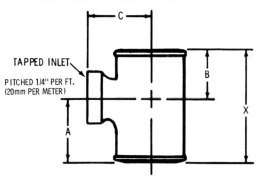

TAPPED INLET
PITCHED 1/4" PER FT.
(20mm PER METER)

SINGLE & DOUBLE
SANITARY T-BRANCHES, TAPPED
"METRIC"

Size	"A"	"B"	"C"	"X"
38 × 32 mm	83 mm	62 mm	65 mm	144 mm
38 × 38 mm	83 mm	62 mm	65 mm	144 mm
51 × 32 mm	83 mm	62 mm	71 mm	144 mm
51 × 38 mm	83 mm	62 mm	71 mm	144 mm
51 × 51 mm	95 mm	67 mm	78 mm	162 mm
76 × 32 mm	83 mm	62 mm	84 mm	144 mm
76 × 38 mm	83 mm	62 mm	84 mm	144 mm
76 × 51 mm	95 mm	67 mm	90 mm	162 mm
102 × 32 mm	83 mm	62 mm	97 mm	144 mm
102 × 38 mm	83 mm	62 mm	97 mm	144 mm
102 × 51 mm	95 mm	67 mm	103 mm	162 mm

Cast-Iron No-Hub Fittings

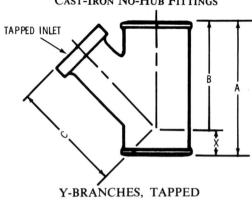

Y-BRANCHES, TAPPED

Size	"A"	"B"	"C"	"X"
2" x 1¼"	6⅝"	5 "	5- 1/16"	1⅝"
2" x 1½"	6⅝"	5 "	5- 1/16"	1⅝"
2" x 2 "	6⅝"	4⅝"	5- 1/16"	2 "
3" x 1¼"	6⅝"	5⅛"	5- 1/16"	1½"
3" x 1½"	6⅝"	5½"	5- 3/4 "	1⅛"
3" x 2 "	6⅝"	5⅛"	5-13/16"	1½"
4" x 1¼"	6⅝"	5⅝"	6- 7/16"	1 "
4" x 1½"	6⅝"	5⅝"	6- 7/16"	1 "
4" x 2 "	6⅝"	5⅝"	6- 1/2 "	1 "

"METRIC"

Size	"A"	"B"	"C"	"X"
51 × 32 mm	168 mm	127 mm	129 mm	41 mm
51 × 38 mm	168 mm	127 mm	129 mm	41 mm
51 × 51 mm	168 mm	117 mm	129 mm	51 mm
76 × 32 mm	168 mm	130 mm	129 mm	38 mm
76 × 38 mm	168 mm	140 mm	146 mm	29 mm
76 × 51 mm	168 mm	130 mm	148 mm	38 mm
102 × 32 mm	168 mm	143 mm	164 mm	25 mm
102 × 38 mm	168 mm	143 mm	164 mm	25 mm
102 × 51 mm	168 mm	143 mm	165 mm	25 mm

Cast-Iron No-Hub Fittings

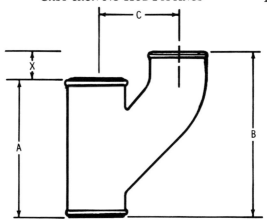

UPRIGHT Y

Size	"A"	"B"	"C"	"X"
2"	7 "	10- 1/4 "	5½"	3- 1/4 "
3"	8⅜"	10- 5/8 "	5½"	2- 1/4 "
4"	9¾"	11- 9/16"	6 "	1-13/16 "
3" x 2"	7 "	9-11/16"	5½"	2-11/16 "
4" x 2"	7 "	9- 1/4 "	5½"	2- 1/4 "
4" x 3"	8⅜"	10- 1/8 "	5½"	1- 3/4 "

"METRIC"

Size	"A"	"B"	"C"	"X"
51 mm	178 mm	260 mm	140 mm	83 mm
76 mm	213 mm	270 mm	140 mm	57 mm
102 mm	248 mm	294 mm	152 mm	46 mm
76 × 51 mm	178 mm	246 mm	140 mm	68 mm
102 × 51 mm	178 mm	235 mm	140 mm	57 mm
102 × 76 mm	213 mm	257 mm	140 mm	44 mm

Cast-Iron No-Hub Fittings

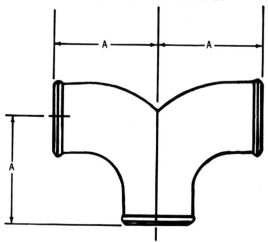

¼ BEND, DOUBLE

Size	"A"
2"	4½"
3"	5 "
4"	5½"

"METRIC"

Size	"A"
51 mm	114 mm
76 mm	127 mm
102 mm	140 mm

Cast-Iron No-Hub Fittings

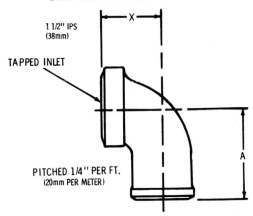

SHORT RADIUS TAPPED ¼ BEND

Size	"A"	"X"
1½" × 1¼"	3"	2"
1½" × 1½"	3"	2"
2" × 1¼"	3¼"	2¼"
2" × 1½"	3¼"	2¼"

SHORT RADIUS TAPPED ¼ BEND "METRIC"

Size	"A"	"X"
38 × 32 mm	76 mm	51 mm
38 × 38 mm	76 mm	51 mm
51 × 32 mm	83 mm	57 mm
51 × 38 mm	83 mm	57 mm

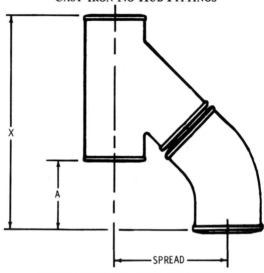

CAST-IRON NO-HUB FITTINGS AND SPECIFICATIONS

Spread between vent and revent when using a WYE, and 1/8th bend, to nearest 1/16".

Size	Spread	"X"	"A"
2" x 2"	5- 5/16"	10- 1/16"	3- 7/16"
3" x 3"	6- 1/4 "	11- 1/2 "	3- 1/2 "
4" x 4"	7- 1/4 "	12-13/16"	3- 5/16"
3" x 2"	5- 3/4 "	10- "	3- 3/8 "
4" x 2"	6- 1/4 "	10 "	3- 3/8 "
4" x 3"	6-13/16"	11- 1/2 "	3- 1/2 "
5" x 2"	7- 5/16"	11 "	2-15/16"
5" x 3"	7- 7/8 "	12- 9/16"	2- 7/8 "
5" x 4"	8- 5/16"	13- 7/8 "	2-11/16"
6" x 2"	7- 7/8 "	11- 1/8 "	2-13/16"
6" x 3"	8- 3/8 "	12- 5/8 "	2- 7/8 "
6" x 4"	8-13/16"	13- 7/8 "	2-11/16"

Cast-Iron No-Hub Fittings

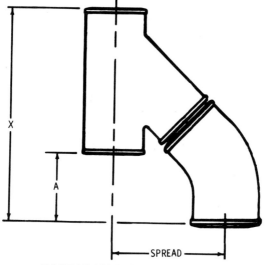

CAST-IRON NO-HUB FITTINGS AND SPECIFICATIONS

Spread between Vent and Revent when using a Wye and ⅛th bend, to nearest millimeter.

"METRIC"

Size	Spread	"X"	"A"
51 × 51 mm	135 mm	256 mm	87 mm
76 × 76 mm	159 mm	292 mm	89 mm
102 × 102 mm	184 mm	325 mm	84 mm
76 × 51 mm	146 mm	254 mm	86 mm
102 × 51 mm	159 mm	254 mm	86 mm
102 × 76 mm	173 mm	292 mm	89 mm
127 × 51 mm	186 mm	279 mm	75 mm
127 × 76 mm	200 mm	319 mm	73 mm
127 × 102 mm	211 mm	352 mm	68 mm
152 × 51 mm	200 mm	283 mm	71 mm
152 × 76 mm	213 mm	321 mm	73 mm
152 × 102 mm	224 mm	352 mm	68 mm

Cast-Iron No-Hub Fittings

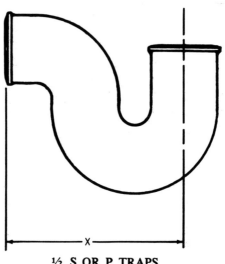

½ S OR P TRAPS

Size	"X"
1½"	6¾"
2 "	7½"
3 "	9 "
4 "	10½"
6 "	14"

½ S OR P TRAPS "METRIC"

Size	"X"
38 mm	171 mm
51 mm	191 mm
76 mm	229 mm
102 mm	267 mm
152 mm	356 mm

CAST-IRON NO-HUB FITTINGS

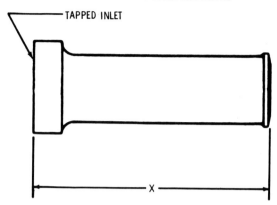

TAPPED EXTENSION PIECE

Size	"X"	"I.P.S. Tapping"
2"	12"	2 "
3"	12"	3 "
4"	12"	3½"

TAPPED EXTENSION PIECE "METRIC"

Size	"X"	"I.P.S. Tapping"
51 mm	305 mm	51 mm
76 mm	305 mm	76 mm
102 mm	305 mm	89 mm

Cast-Iron No-Hub Fittings

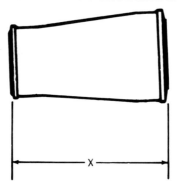

INCREASER - REDUCER

Size	"X"
1½" x 2"	3⅝"
2" x 3"	8"
2" x 4"	8"
3" x 4"	8"
4" x 6"	4"
5" x 6"	4½"

INCREASER - REDUCER "METRIC"

Size	"X"
38 × 51 mm	92 mm
51 × 76 mm	203 mm
51 × 102 mm	203 mm
76 × 102 mm	203 mm
102 × 152 mm	102 mm
127 × 152 mm	114 mm

Cast-Iron No-Hub Fittings

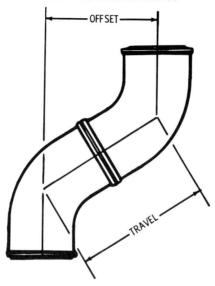

TABLE 1
MINIMUM OFFSETS USING NO-HUB
⅛th BEND FITTINGS

Size	Travel	Min. Offset
1½"	5-11/32"	3-13/16"
2 "	5-19/32"	3-15/16"
3 "	6- 3/32"	4- 5/16"
4 "	6-11/32"	4- 1/2 "
5 "	7-27/32"	5- 9/16"
6 "	8- 7/32"	5-13/16"

TABLE 2
MINIMUM OFFSETS USING NO-HUB 1/16th BEND FITTINGS

Size	Travel	Min. Offset
2"	4-11/32"	1-11/16"
3"	4-19/32"	1- 3/4 "
4"	4-23/32"	1-13/16"
5"	5-31/32"	2- 5/16"
6"	6- 3/32"	2- 5/16"

TABLE 3
MINIMUM OFFSETS USING NO-HUB 1/6th BEND FITTINGS

Size	Travel	Min. Offset
2"	6-19/32"	5-11/16"
3"	7- 3/32"	6- 1/8 "
4"	7-23/32"	6-11/16"
5"		
6"		

TABLE 4
MINIMUM OFFSETS USING NO-HUB 1/5th BEND FITTINGS

Size	Travel	Min. Offset
2"	7-15/32"	7- 1/16"
3"	8- 7/32"	7-13/16"
4"	8-31/32"	8- 1/2 "
5"		
6"		

Note: Minimum Offsets figured to nearest 1/16".

Cast-Iron No-Hub Fittings

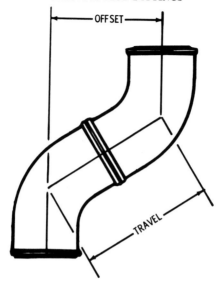

TABLE 1
MINIMUM OFFSETS USING NO-HUB
⅛th BEND FITTINGS
"METRIC"

Size	Travel	Min. Offset
38 mm	136 mm	97 mm
51 mm	142 mm	100 mm
76 mm	155 mm	110 mm
102 mm	161 mm	114 mm
127 mm	199 mm	141 mm
152 mm	209 mm	148 mm

TABLE 2
MINIMUM OFFSETS USING NO-HUB
1/16th BEND FITTINGS
"METRIC"

Size	Travel	Min. Offset
51 mm	110 mm	43 mm
76 mm	117 mm	44 mm
102 mm	120 mm	46 mm
127 mm	152 mm	59 mm
152 mm	155 mm	59 mm

TABLE 3
MINIMUM OFFSETS USING NO-HUB
1/6th BEND FITTINGS
"METRIC"

Size	Travel	Min. Offset
51 mm	167 mm	144 mm
76 mm	180 mm	156 mm
102 mm	196 mm	170 mm

TABLE 4
MINIMUM OFFSETS USING NO-HUB
1/5th BEND FITTINGS
"METRIC"

Size	Travels	Min. Offset
51 mm	190 mm	179 mm
76 mm	209 mm	198 mm
102 mm	228 mm	216 mm

Note: Minimum Offsets figured to nearest millimeter.

Cast-Iron No-Hub Fittings 129

Note sway brace and method of hanging.

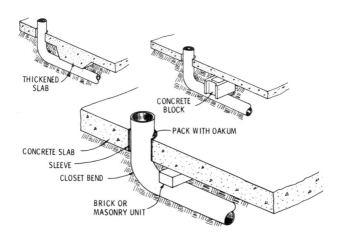

130 CAST-IRON NO-HUB FITTINGS

Note sway brace, method of hanging and cleanout.

Note neat method of hanging pipe.

Cast-Iron No-Hub Fittings 131

*Method of using hanger for closet bend.
Note sleeves and oakum in sleeve.*

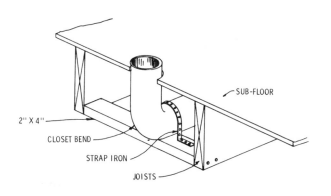

Bracing for Closet Bend

132   Cast-Iron No-Hub Fittings

View showing method of hanging and sway brace.

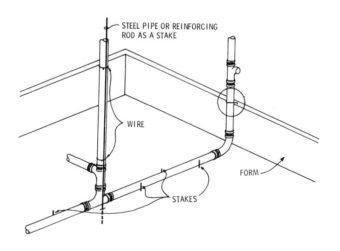

Slab-on-Grade Installation

Cast-Iron No-Hub Fittings 133

Method of clamping the ℄ NO-HUB Pipe at each floor, using a friction clamp or floor clamp.

134   CAST-IRON NO-HUB FITTINGS

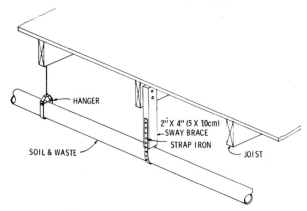

Horizontal Pipe with Sway Brace

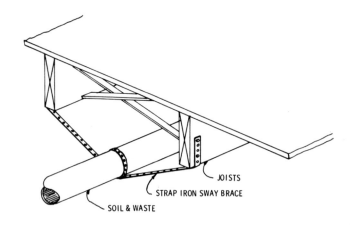

Sway Brace

Cast-Iron No-Hub Fittings 135

WIRE STAPLE

FORM OR SILL

PIPE ON GRADE

Support for Vertical Pipe

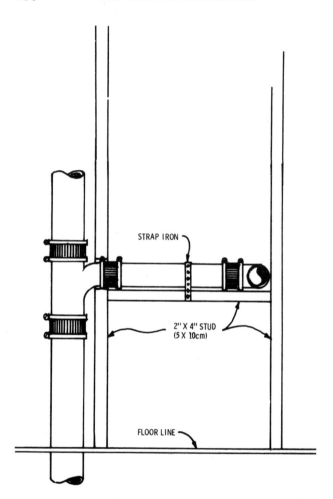

SILVER BRAZING AND SOFT SOLDERING

I wish to thank NIBCO, Inc., Elkhart, Indiana, for granting permission to reprint certain portions of their copper piping manual relating to silver brazing and soft soldering. The following information on silver brazing & soft soldering has been taken from (NIBCO CPM-1) copper piping manual.

Flux passes through four stages:

1. At 212°F. (100°C.) the water boils off.
2. At 600°F. (316°C.) the flux becomes white and slightly puffy and starts to work.
3. At 800°F. (427°C.) it lays against the surface and has a milky appearance.
4. At 1100°F. (593°C.) it is completely clear and active and has the appearance of water.

SILVER BRAZING INFORMATION

Size of Copper Tube		Oxygen—Pressure P.S.I.—kPa		Acetylene—Pressure P.S.I.—kPa	
½" & ¾"	13 & 19 mm.	5	34.5	5	34.5
1" & 1¼"	25 & 32 mm	6	41.4	6	41.4
1½", 2" & 2½"	38, 51 & 64 mm	7	48.26	7	48.26
3" & 3½"	76 & 89 mm	7½	51.7	7½	51.7
4", 5" & 6"	102, 127 & 152 mm	9	62	9	62

SILVER BRAZING
"FLUX"

Note: Avoid excess flux, and avoid flux on areas not cleaned. Particularly avoid getting flux into the inside of the tube itself.

Apply *Heat and Brazing* Alloy:

The preferred method is by the oxy-acetylene flame. Propane and other gases are sometimes used on smaller sizes.

A slightly reducing flame should be used, with a slight feather on the inner blue cone, the outer portion of the flame is pale green. Heat the tube first, beginning at about one inch from edge of the fitting. Sweep the flame around the tube in short strokes up and down at right angles to the run of the tube. It is very important that the flame be in continuous motion and should not be allowed to remain on any one point to avoid burning through the tube.

Generally the flux may be used as a guide as to how long to heat the tube; continue heating after the flux starts to bubble or work and until the flux becomes quiet and transparent, like clear water.

Now switch the flame to the fitting at the base of the cup. Heat uniformly, sweeping the flame from fitting to tube until the flux on the fitting becomes quiet. Particularly avoid excessive heating of cast fittings.

When the flux becomes liquid and transparent on both the tube and the fitting, start sweeping the flame back and forth along the axis of the joint to maintain heat on the parts to be joined; Especially toward the base

SILVER BRAZING

of the cup of the fitting. The flame must be kept moving to avoid burning the tube or fitting.

When joint has reached proper temperature, apply brazing wire or rod. Keep the flame away from the rod or wire as it is fed into joint.

Keep both the fitting and the tube heated by moving the flame back and forth from one to the other as the alloy is drawn into the joint.

When joint is filled, continuous fillet of brazing alloy will be visible completely around joint. Stop feeding as soon as joint is filled. Note: For larger size tube, 1" (25 mm) and larger, it is difficult to bring the whole joint up to heat at one time.

If difficulty is encountered in getting the entire joint up to the desired temperature, a portion of the joint can be heated and brazed at a time. At the proper brazing temperature the alloy is fed into the joint and the torch is then moved to an adjacent area and the operation carried on progressively all around the joint taking care to overlap each operation.

Horizontal and Vertical Joints

When making horizontal joints it is preferable to start applying the brazing alloy at the top, then the two sides, and finally the bottom, making sure that the operations overlap.

On vertical joints, it is immaterial where the start is made. If the opening of the socket is pointed down, care should be taken to avoid overheating the tube, as this may cause the alloy to run down the tube. If this condition is encountered, take the heat away and allow the alloy to set; then reheat the band of the fitting to draw up the alloy.

After the brazing alloy has set, clean off remaining flux with a wet brush or swab.

Wrought fittings may be chilled quickly. However it is advisable to allow cast fittings to cool naturally to some extent before applying a swab.

If the brazing alloy refuses to enter the joint and tends to flow over the outside of either member of the joint, it indicates this member is overheated, or the other is underheated, or both.

If the alloy fails to flow, or has a tendency to ball up, it indicates oxidation on the metal surfaces, or insufficient heat on the parts to be joined.

MAKING UP A JOINT

The preliminary steps of tube measuring, cutting, burr removing and cleaning. (Tube ends and sockets must be thoroughly cleaned before beginning brazing operation.) are identical to the same steps in the "Soft Soldering Process."

A suitable flux can be made which will be suitable for making silver solder joints on copper tubing by mixing powdered Borax and alcohol or water to a thin milky solution.

Soft Soldering

1. Avoid pointing flame into socket opening of fitting.

2. Never apply flame directly on solder.

3. On tubing 1" (25 mm) and above a mild heating of tube before playing flame on fitting is recommended. This greatly enhances a better made joint. Assuring solder to be drawn into the joint by the natural force of cappillary attraction.

4. Flame should be played at base of fitting, flame pointing in the direction of socket opening. This assures any impurities, including excess flux will be flushed out ahead of solder as joint is filled.

If flame were pointed towards base of fitting there would be a chance of these impurities or flux to be trapped inside of joint, creating a flux pocket.

This flux pocket prevents solder from completely occupying inside of socket.

5. When the metal is hot enough, the flame should be moved away and the solder applied.

6. On larger size tubing it is best to hold flame long enough at base of fitting and moved around circumference, to assure evenly distributed heat and solder occupying entire space.

7. As joint cools to lower temperature, continue to apply solder around entire face of fitting; this will create a fillet which assures a full joint. Heat rises and sometimes the top part of a horizontal joint is too hot to retain solder allowing it to run out; or in a tight inaccessible place a portion may have not been heated, this can be detected

by running solder around entire face, if it runs smoothly around creating a fillet a better guarantee is received that a good joint has been made, more assurance added.

8. On horizontal joints it is recommended that flame be played at bottom of fitting and solder applied at the top, however this is merely preferred by the majority of plumbers.

PURPOSE OF FLUX

Purpose of flux is to dissolve residual traces of oxides, prevent oxides from forming during heating, and to float out oxides ahead of solder.

ILLUSTRATIONS ON BRAZING AND SOLDERING

The following illustrations have been provided by the Copper Development Association Inc., 405 Lexington Ave.,

1 Fluxing

2 Assembling

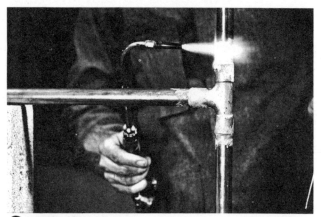

3 Heating Tube

4 Heating Large Tube

ILLUSTRATIONS ON BRAZING

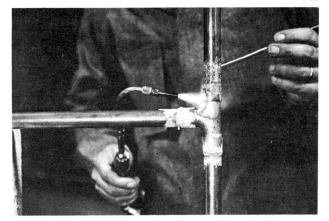

5 Heating Fitting

6 Heating Large Fitting

ILLUSTRATIONS ON BRAZING

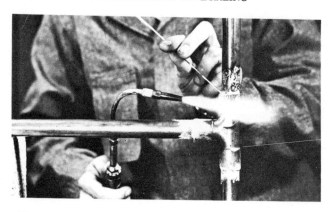

7 Feeding Brazing Alloy

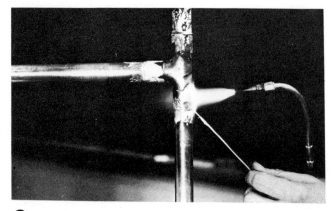

8 Feeding Upward

ILLUSTRATIONS ON BRAZING

9 Swabbing

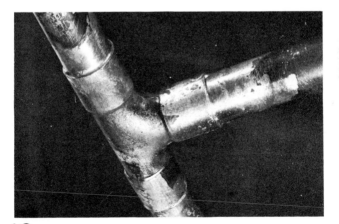

10 Completed Joint

148 Illustrations on Soft Soldering

Removing Burrs

Cleaning Tube End

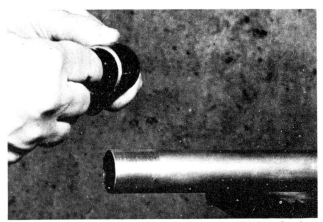

Cleaning Fitting Socket

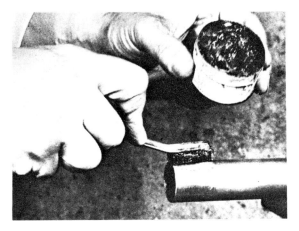

Fluxing Tube End

150   ILLUSTRATIONS ON SOFT SOLDERING

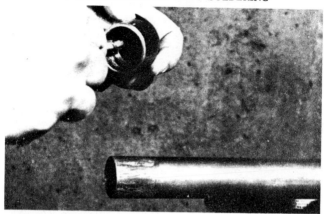

Fluxing Fitting Socket

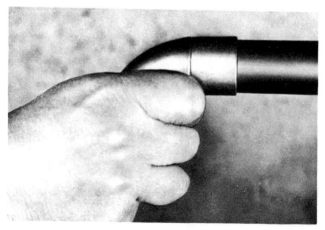

Assembling

ILLUSTRATIONS ON SOFT SOLDERING

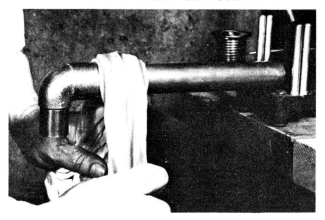

Removing Surplus Flux

Heating

152 ILLUSTRATIONS ON SOFT SOLDERING

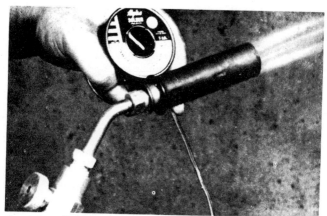

Applying Solder

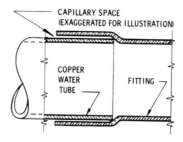

LEAD WORK

I wish to thank Lead Industries, 292 Madison Ave., New York, N. Y. 10017, for making the following section on Lead Work possible. I am very proud to include this section, taken from "Lead Work for Modern Plumbing" copyright 1952, in this new Plumber's Handbook.

A plumber should buy only good wiping solder that has the manufacturer's name and brand indicating composition cast on the bar. Refusal to accept anything less will eliminate many solder troubles that can and does occur.

Good wiping solder usually contains between 37 and 40% pure tin, between 63 and 60% pure lead.

Complete solidification around 360°F. (182°C.); complete liquefaction around 460°F. (238°C.).

In general, the working range will be about 100°F. (38°C.).

Many plumbers twist a piece of newspaper and dip it into molten solder, which is at wiping heat when the paper scorches but does not ignite.

"Preparing Horizontal Round Joints."

First, the ends of the pipe to be joined should be squared off with a coarse file or rasp and the pipe should be drifted so it is uniformly round and free from dents.

Grease and dirt should be cleaned off the surface of the pipe for about 4" (102 mm) from the ends. With a knife, ream out burrs in the ends of pipe to be joined.

Then one end should be flared, using a turnpin and mallet, until the inside diameter at the end equals the original outside diameter. The end thus flared being the one in which the water will flow. The shoulder or outside edge of the flare should then be rasped off approximately parrallel with outside wall of the pipe .

An adequate flare for lead pipe is ¼" (6.4 mm) to ⅜" (9.5 mm).

The inside of the flared end should be soiled for about 1" (25 mm). Next, the end of the other section of pipe, which will be the one from which the water will flow, should be beveled with a rasp until it fits snugly inside the flared end. In this way, joint is made in direction of flow, reducing resistance and chance of clogging. With dividers, mark a line around the flared end of pipe at a distance from the extreme end equal to half the length of the finished joint (half being generally 1¼" or 32 mm).

Mark a similar line on the beveled end at a distance from the extreme end equal to half the length of the joint, plus the length of the beveling. This will make the center of the finished joint at the intersection of the two outside surfaces of pipe. Next mark an additional 3" (76 mm); in this 3" (76 mm) portion "being at extreme ends of both pieces of pipe to be joined," clean lightly with wire brush, dust off and apply plumber's soil.

Next the 1¼" (32 mm) portions on each pipe totaling 2½" (64 mm) joint, the flare and bevel should be lightly scraped clean with a shave hook and immediately covered with a thin coating of tallow to prevent oxidation.

Now the ends of pipe should be fitted snugly together and braced in position so they will be absolutely stationary during and after wiping until the solder has cooled.

LEAD WORK

The bottom of the pipe should be about 6" (152 mm) above bench or working place. When wiping joints in place, if the space is more than 6" (152 mm), a box or some other flat object should be placed so there will be a surface about 6" (152 mm) under the joint to prevent splashing of solder.

To aid in getting heat up on a joint quickly, one or both of the outer, or extreme ends of the pipe should be plugged, usually newspaper is used.

After wiping solder is heated, carefully stirred and skimmed to remove dross, testing for proper heat (as described previously), which is 600°F. (316°C.), wiping may begin.

Wiping cloths are usually made of 10 oz. (284 gram) herring bone material.

The generally used wiping cloth sizes are: 3" (76 mm) cloth for 2½" (64 mm) joint; 3¼" (83 mm) cloth for 2¾" (70 mm); measured by length of joints.

PROCEDURE FOR CLEANING WIPING SOLDER

The procedure for cleaning wiping solder varies, one method is: Heat solder to a dull red, about 790° F. (421° C.) (melting point of zinc), then add about one tablespoon of sulphur and stir; then permit pot to cool slowly and skim off top dross which contains the impurities consisting of compounds of lead, tin, zinc. Stir and skim until top is clean. Next, add a small amount of powdered or lump rosin, stir and skim again.

Allow pot to cool until solder reaches wiping temperature; Then add sufficient tin to re-establish proper workability.

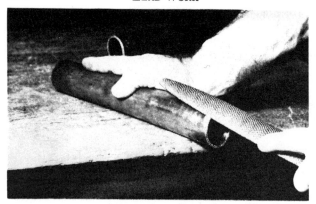

Beveling male end of lead pipe with a rasp preparatory to fitting it into the flared end for wiping. A close fit is highly important to successful joint wiping.

Scraping male end clean after beveling and soiling. Female end, at left, flared, soiled and cleaned, ready to have the male end inserted in it.

Lead Work

Pipes prepared and fitted together ready for wiping. Note how they are held securely by boards and bricks with small boards under one side to prevent rolling. The paper under the joint is to catch excess solder.

Solder has just been poured on the joint and that caught in the cloth is being pressed against the bottom to get up heat.

Excess solder has been removed from the soiling and shaping started across the bottom and up the side next to the wiper.

The hand has been reversed and the stroke continued across the top and down the side away from the wiper.

A handy method of retaining heat under adverse conditions by building up solder a little way from each side of the joint to be lifted off after completion.

PREPARATION TO WIPING 4″ (102 MM) HEAD STUB ON JOB BRASS FERRULE

Prepare by using fine file, apply Nokorode, proceed to tin ferrule.

LEAD STUB

Use fine file to remove rough edges, use knife to ream; with flat dresser proceed to work lead till it fits snugly into brass ferrule (assume this type ferrule is used). Approximately 1⅛″ (29 mm) to 1¼″ (32 mm) should be inserted into ferrule. Next, use shave hook, shave a portion equal to 1″ (25 mm) + insertion say 2¼″ (57 mm), apply mutton tallow. Then insert into ferrule, proceed to secure; then apply a ring of plumber's soil inside ferrule brass side where lead ends inside.

Next, use dividers mark 1″ (25 mm) line around lead from face of ferrule; next use wire brush to clean additional 3″ (76 mm) of lead. Apply soil on this portion, try not to apply to 1″ (25 mm) portion being readied to receive solder. Next, use shave hook to clean 1″ (25 mm) portion. Then reapply mutton tallow. Last, apply gummed paper to ferrule.

After wiping, prepare lead cap, readying stub for testing.

4″ (102mm) LEAD STUB

- SOLDERED CAP
- PLUMBER'S SOIL
- 2″ (51mm)
- GUMMED PAPER
- SOIL INSIDE AND OUTSIDE OF EXPOSED PORTION OF FURREL
- SOLDERED JOINT

NOTE: THIS METHOD IS ASSUMING PLUMBER'S SKILL AND NOT PRODUCTION METHOD.

LEAD WORK

LEAD JOINING WORK

In the regular run of his lead-joining work, the plumber usually needs a pair of two-lb. (907 gram) soldering irons.

For ordinary work, the point of the soldering iron should be tinned on all four sides for a distance of at least ¾" (19 mm) from the tip. If, however, soldering is to be done from underneath a joint, the iron should be tinned on one surface only. The side to be used next to the joint being soldered. This permits control of the solder and prevents it from running away from the joint.

LAP JOINT

A lap of about ⅜" (10 mm) is advisable for the weights of lead ordinarily employed by plumbers. On top surface of bottom sheet, a line should be marked ½" (12.7 mm) back from the edge to be joined. This portion shaved lightly with the shave hook, using strokes parallel to the edge, until the surface is clean back to the line.

The same procedure should be followed on the under side of the top sheet. The edge of top sheet should also be cleaned and the top sheet placed in position lapping over the other ⅜" (10 mm). This leaves ⅛" (3.2 mm) of the cleaned portion of the lower piece of lead exposed. With the flat dresser, the top sheet should be dressed down to the level of the bottom sheet except where it actually laps and is held up by the lead underneath. The lap should be dressed to fit snugly. With the shave hook, the upper surface of the top layer should then be cleaned for a distance of ⅜" (10 mm) from the edge. Tallow should be applied immediately to all cleaned areas. As in making butt joints, the sheets should be tacked in similar manner.

"BUTT JOINT"

To make a butt joint, edges to be joined should be beveled with a shave hook so that they make an angle of 45° or more with the vertical.

This is accomplished by wrapping a piece of cloth around index finger of the hand holding shave hook and using this finger, pressed against the edge of the lead, as a guide when drawing the shave hook along the edge.

Immediately after shaving, tallow candle or refined mutton tallow free from salt should be rubbed over all shaved parts in a very thin coat to prevent oxidation.

Edges to be joined should then be placed firmly together and powdered rosin sprinkled along the joint. With a clean, well tinned soldering iron and 50-50 solder, the edges are next tacked together at intervals of from 4" (102 mm) to 6" (152 mm), using a drop of solder at each point. An iron at proper heat should then be placed against lead at the end of the seam in the groove formed by the abutting beveled edges.

50-50 solder should be fed in slowly allowing it to be melted by the iron and fill the groove. The iron should be drawn slowly along the joint, the speed being such as to permit the solder to melt and fill the groove continuously, building up to a slightly rounded surface when finished.

ROUGHING AND REPAIR INFORMATION
As a General Rule
and if Roughing Sheets not handy

LAVATORIES

Note: Hot and cold water will rough at 20½" (521 mm) if using speedy supplies; if brass nipples are used check your roughing in sheets.

Waste at 17½" (444 mm) will apply to pop-up waste, or drain plug.

Backing C/L of backing for bracket 31" (787 mm) using a 2" × 10" (5 × 25 cm).

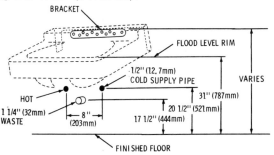

3/8" (9.5mm) SUPPLY TO FIXTURE

CORNER LAVATORIES

Waste: 17½" (444 mm) from finished floor, 6¾" (171 mm) from corner to center (left or right).

Water: Using speedy supplies 20½" (521 mm) from finished floor, 7" (178 mm) from corner to center; hot on left, cold on right.

Backing: C/L of backing 32" (813 mm) using a 2" × 8" (5 × 20 cm).

VANITY LAVATORIES

Waste: Not over 16" (406 mm) from finished floor.
Water: Not over 18" (457 mm) from finished floor.

WATER CLOSETS

Flush Valve: 4¾" (120 mm) to right from center line.

Cold Water Supply: 1" (25.4 mm) S.P.S. 26" (660 mm) from finished floor or 20½" (521 mm) from center of wall hung waste to center of water.

KITCHEN SINK

Waste: 1½" (38 mm)—22¼" (565 mm) from finished floor, 8" (203 mm) off center line of a double compartment sink. Single compartment sink will rough at 25¼" (641 mm).

Hot and cold water will rough in at 23" (584 mm) from finished floor; hot 4" (102 mm) to left of center line, cold 4" (102 mm) to right of center line.

Note: If one compartment of a two compartment sink is to be provided with a garbage disposal, rough the waste at 16" (406 mm) above finished floor.

Service sinks rough in at 10½" (267 mm) on waste.

Water generally is roughed at 6" (152 mm) from finished floor; hot 4" (102 mm) to left, cold 4" (102 mm) to right.

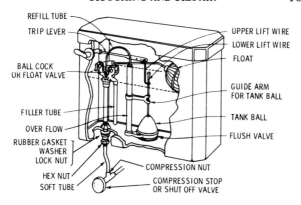

WATER CLOSET FLUSH TANK

NOTE:

To replace a ball cock or float valve

1. Close valve that supplies water to tank.
2. Flush tank and remove remaining water with sponge or rag.
3. Holding float valve with one hand to prevent its turning, begin loosening hex nut, or nut securing supply tube to float valve.
4. Then begin loosening lock nut, lift float value out of tank.
5. Before placing new float value in tank be sure the spot is clean, and free of dirt or rust.
6. Follow instruction in reverse order, be sure and use pipe dope on threads.
7. When new float valve is installed take refill tube, screw it into opening provided, hold refill tube at one end, then bend other end until it enters into overflow pipe.

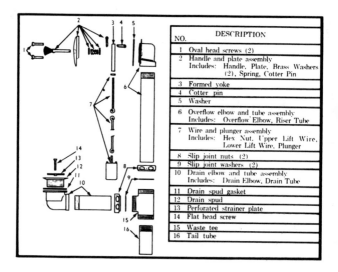

NO.	DESCRIPTION
1	Oval head screws (2)
2	Handle and plate assembly Includes: Handle, Plate, Brass Washers (2), Spring, Cotter Pin
3	Formed yoke
4	Cotter pin
5	Washer
6	Overflow elbow and tube assembly Includes: Overflow Elbow, Riser Tube
7	Wire and plunger assembly Includes: Hex Nut, Upper Lift Wire, Lower Lift Wire, Plunger
8	Slip joint nuts (2)
9	Slip joint washers (2)
10	Drain elbow and tube assembly Includes: Drain Elbow, Drain Tube
11	Drain spud gasket
12	Drain spud
13	Perforated strainer plate
14	Flat head screw
15	Waste tee
16	Tail tube

TYPICAL TUB-TRIP WASTE AND OVERFLOW INSTALLATION INSTRUCTIONS

1. Remove flat head screw (14) and perforated strainer plate (13) from drain spud (12). Apply small amount of putty to underside of drain spud.

2. Insert drain spud (12) through tub drain outlet from the inside; place drain spud gasket (11) on face of drain elbow as shown. Proceed to tighten drain spud (12) until it is secure. When spud (12) is secure, drain tube should point directly to the end of the tub. Replace

perforated strainer plate and flat head screw into drain spud.

3. Place slip joint nut and one slip joint washer on the drain tube or shoe (10), and one slip joint nut and one slip joint washer, on riser tube (6). Place riser tube into the long end of the waste tee (15) and hand tighten slip joint nut. Place washer (5) on the face of the overflow elbow and push the complete assembly onto the drain tube and hand tighten slip joint nut.

4. Line up washer with the overflow opening in the tub. Insert plunger and wire inside of the tub and feed plunger and wire through opening until handle and plate lines up with the overflow opening in the tub. Secure plate in place by screwing two oval head screws through plate into overflow elbow.

5. Wrench tighten two slip joint nuts so that the drain tube and the riser tube are sealed to the waste tee.

6. The tub may now be placed in position with the tail tube or tailpiece (16) slipped into the drainage line or connected to — and sealed tight.

Note: Depending on size of tub, occasionally drain tube (10) and overflow tube (6) will need to be cut shorter.

Note: Wire and plunger assembly may need to be adjusted so that drain will work properly.

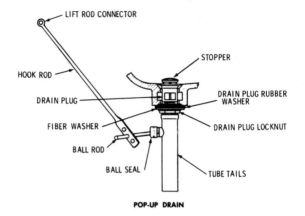

POP-UP DRAIN

TYPICAL POP-UP DRAIN
To Install Pop-Up Drain

1. Remove drain plug from tube tail so as to detach locknut, rubber washer and fiber washer.
2. Insert drain plug through drain hole of lavatory using plumber's putty underneath flange of plug. Attach rubber washer, fiber, or metal washer and locknut.
3. Assemble tube tail to drain plug using a good pipe joint compound and tighten. Turn drain so that side hole in tube is pointed to the rear of lavatory. Now tighten locknut.
4. Assemble ball rod assembly to hole in side of tube. Tighten loosely by hand.
5. Insert stopper into drain.
6. Attach hook rod as shown and tighten set screw so that drain works properly by operating knob.

ROUGHING AND REPAIR
BATHROOM ILLUSTRATION

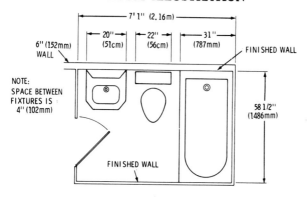

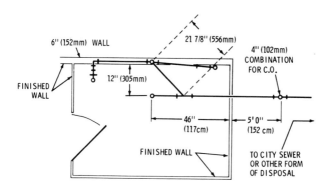

Note: Check Local Code.

BATHROOM ILLUSTRATION

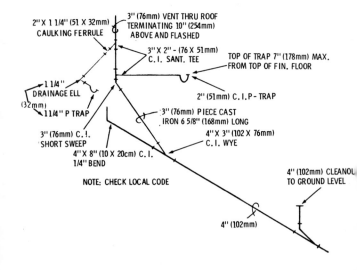

Note: Above Illustration is using cast-iron pipe and fittings with hubs.

Roughing and Repair

"DOMESTIC WASHING MACHINE ROUGH-IN INFORMATION"

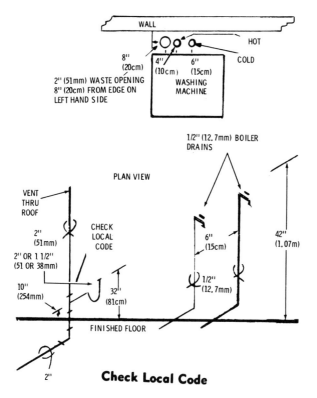

Check Local Code

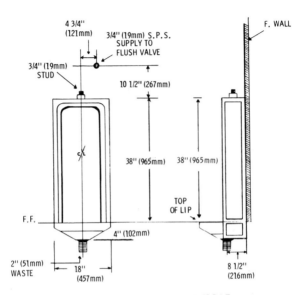

TYPICAL STALL URINAL

Note: Opening left for urinal should be approx. 24" (61 cm) wide and 18½" (47 cm) from finished wall.

Note: 2" (51 mm) waste should be ¼" (7 mm) to ⅜" (10 mm) below top of spud or strainer.

Note: Top of lip should be ¼" (7 mm) below finished floor, some codes call for top of lip to be above finished floor.

Check Local Code.

Note: Sharp sand should be packed under urinal; and when urinal is set, sand should be packed at least 1 inch (25 mm) up on base of urinal.

When urinals are placed in a battery, spreaders are available; 3" (76 mm) spreaders are popular.

ROUGHING AND REPAIR

REPAIRING WATER FAUCETS AND VALVES
KITCHEN FAUCET OR LAVATORY:

Refer to "Top Mount Sink Fitting" under faucets, for part description.

1. Shut off water supply valves or stops .
2. Remove handle screw (1).
3. Remove handle (2); be sure faucet is not completely closed before attempting to loosen lock nut (3), or on many faucets packing nut, that will allow you to remove stem (5). Open faucet ¼ turn and continue to check as lock nut is loosened.

A crescent wrench should be used.

4. Stem (5) can then be removed.
5. Replace washer at bottom of stem.

Note: If Bibb screw holding washer appears old and difficult to remove, cut washer out with a penknife. A pair of pliers can them be used, whereas before a screwdriver may have turned off part of head.

6. Before replacing stem, examine the seat (10) located at bottom (inside valve body) where washer seats. If seat (10) appears rough, or a notch or groove discovered, seat should be replaced.

This seat can be removed by using an Allen wrench in most cases.

Note: If seat is not too badly worn, and if seat is of the unremovable type, it can be refaced by using a "seat dressing tool".

Note: Cause of badly worn seats in most cases is delay in replacing worn washers; the passing of water "when valve is shut" between washer and seat causes a notch or groove to be worn.

7. When stem is again inserted back into faucet body, remember to make sure it's kept open slightly. This will prevent damage to stem.

ROUGHING AND REPAIR

Note: If water leaks out of handle, it's caused by a worn "O" ring; a thin rubber ring located on the stem.

On some faucets; and on valves such as the "globe valve" illustrated here, the packing nut is the cause of leakage.

Open valve ¼ turn, tighten nut snugly; if valve continues to leak, a new packing washer must be installed; you may also wrap stranded graphite packing around spindle and tighten snugly.

Note: If spout (15) leaks where it enters body of valve, this is also caused by worn out "O" ring.

Note: The "American Standard" Aquaseal Valve has a diaphragm (as illustrated in the Aquaseal Kit shown in Faucets) in place of a washer.

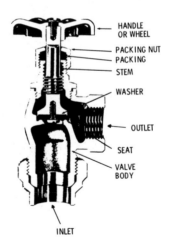

Note: Some faucets are similar in design to the above globe valve.

Roughing and Repair

Faucets

I wish to thank American Standard, P. O. Box 2003, New Brunswick, New Jersey, for providing the illustrations and data on Faucets.

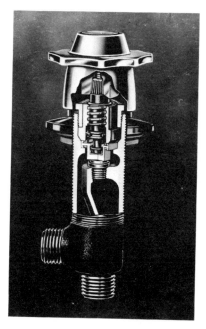

NO-DRIP AQUASEAL VALVE

The secret behind the no-drip feature offered by American-Standard Hermitage™ fittings lies in this washerless valve. All moving parts are outside the flow area, and lubrication on the stem threads is effective for the life of the fitting. There is no seat washer wear — the cause of leaks and dripping in ordinary fittings.

ROUGHING AND REPAIR

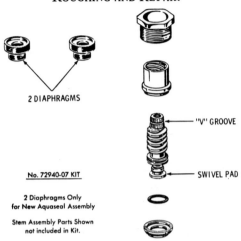

2 DIAPHRAGMS

No. 72940-07 KIT

2 Diaphragms Only
for New Aquaseal Assembly

Stem Assembly Parts Shown
not included in Kit.

"V" GROOVE

SWIVEL PAD

IMPORTANT

When it becomes necessary to replace the diaphragm in your aquaseal valve trim, remove the handle and check for a "V" groove around stem located in the middle of the splines.

If you DO NOT SEE the "V" groove, replace the handle and contact your supplier for the appropriate stem assembly.

If the "V" groove is visible, proceed to remove the valve unit.

After removing the old diaphragm turn the stem so that one thread still protrudes from the top of the stem nut. Slip new diaphragm over the swivel pad and insert assembly into the fitting exercising normal care to prevent damaging the diaphragm.

Tighten valve unit and replace handle.

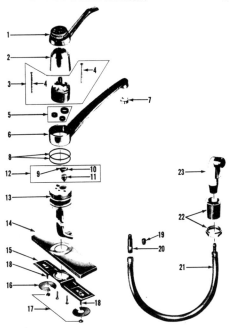

SINGLE CONTROL AQUARIAN SINK FITTING
Faucet as Shown with Numbers

Description

1. Handle
2. Escutcheon Cap
3. Cartridge (4 Gal.)
4. Cartridge Screw
5. Cartridge Seal Set
6. Spout S/A W/Aerator
7. Aerator
8. "O" Ring
9. Divertor
10. "O" Ring
11. Retainer S-A
12. Diverter & Retainer Set
13. Manifold (4 Gal.)
14. Escutcheon
15. Mounting Plate
16. Washer Slotted
17. Nut
18. Carriage Bolt
19. Pipe Plug
20. Hose Connection
21. Hose S/A
22. Locknut & Spray Holder
23. Spray Head S/A

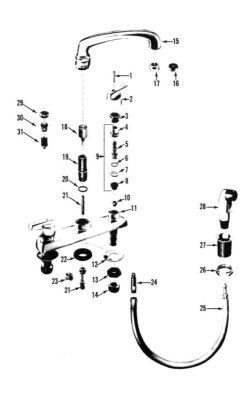

**TOP MOUNT SINK FITTING
HERMITAGE TRIM — AQUASEAL**
Illustration of Faucet

Roughing and Repair 179

Description

1. Handle Screw
2. Handle
3. Lock Nut
4. Stem Nut
5. Stem w/Swivel
6. Friction Ring
7. Stop Ring
8. Aquaseal Diaphragm
9. Aquaseal Trim
10. Seat
11. Body
12. Friction Washer
13. Lock Nut
14. Coupling Nut
15. Spout
16. End Trim

17. Aerator
18. Divertor
19. Post
20. "O" Ring
21. Hose Connection Tube
22. Gasket
23. Body Plug
24. Hose Connector
25. Hose S/A
26. Lock Nut
27. Spray Holder
28. Spray Head
29. Cap w/Washer
30. Auto Spray (Upper)
31. Auto Spray (Lower)

Note: If part #18 is not used, order parts 29, 30 & 31.

WORKING DRAWINGS

FRESH AIR SYSTEM

Installed in places where food is sold, stored, handled, manufactured, or processed. Such as restaurants, cafes, lunch stands, dairies, bakeries, etc.

Note: Fresh air master trap shall not be less than 4" (102 mm).

Note: Fresh air inlet may extend out building wall approximately 12" (31 cm) above grade or extend thru roof. This inlet and aux. vent shall not be connected to any sanitary vent stack. Check local code.

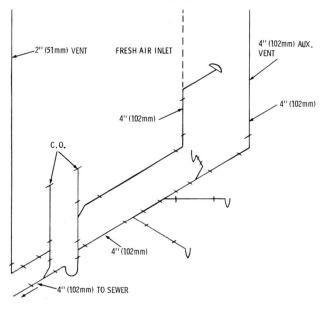

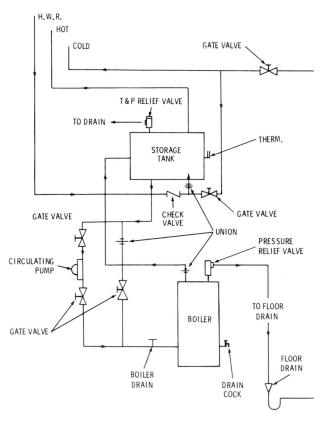

CONNECTIONS BETWEEN HEATER AND STORAGE TANK WITH BY-PASS

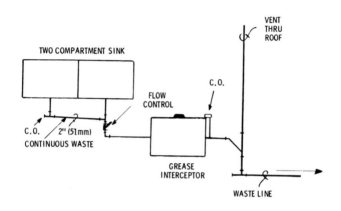

TYPICAL ILLUSTRATION OF A GREASE TRAP

Note: No sink trap is needed when sink is connected to a grease trap. The interceptor is a trap itself and will prevent sewer gases from entering house.

Note: A vent is provided on outlet, or sewer side to prevent siphonage of the contents of the grease trap.

Note: Grease traps should be installed so as to provide access to the cover and means for servicing and maintaining the trap in working and operating condition.

Note: Check local code in your area, and procedure set up by the administrative authority.

WORKING DRAWINGS 183

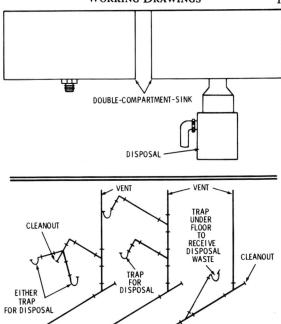

Note: The three illustrations above, show three methods of roughing in the waste for a two compartment sink with disposal.

Note: The first illustration is common in alteration work; a double drainage wye is inserted in the waste line, another waste arm and separate trap to provide for waste of disposal can now be run.

Note: Check local code for proper sizing of waste pipes traps, and vent pipes.

Note: In illustration (2) rough lower sanitary tee at 16" (41 cm) to 18" (46 cm) above finished floor.

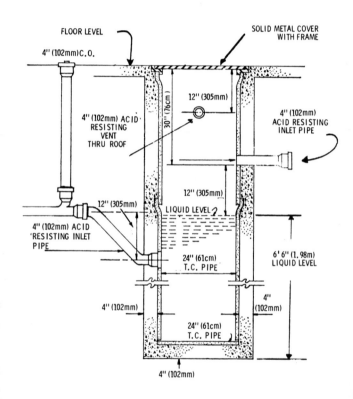

ACID DILUTING TANK

WORKING DRAWINGS

SAND TRAP

Typical Example

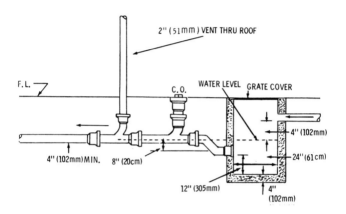

Note: Inside 24″ × 24″ (61 × 61 cm) min. size.

Note: In detail above 33½″ (85 cm) would be bottom of pit to bottom of inlet or invert.

Sand traps are generally used in filling stations, garages, poultry houses, places where water carries sand, loam, refuse or other material which would normally clog ordinary drain.

Note: Where sand trap is located in an open area such as a wash rack, slab or other similar places this (2″) (51 mm) vent may be omitted. Check local code.

ELECTRIC CELLAR DRAIN

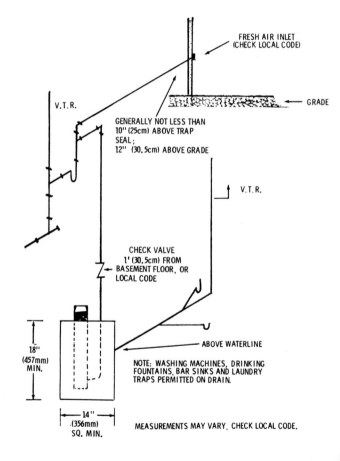

WORKING DRAWINGS
TRAILER CONNECTION ROUGH-IN

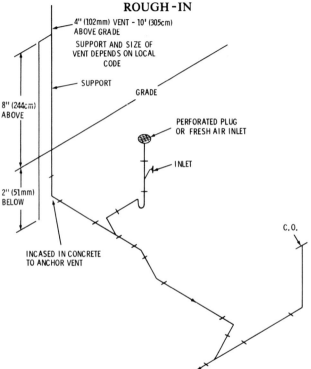

Note: P-trap at least 18" (46 cm) below grade; inlet not more than 4" (102 mm) above grade.

Note: Connection from trailer to inlet should not exceed 8' (244 cm).

Note: The minimum distance between sewer and water connection should be 5' (152 cm). Check local code.

WORKING DRAWINGS
½ S OR P-TRAP AND ITS PARTS

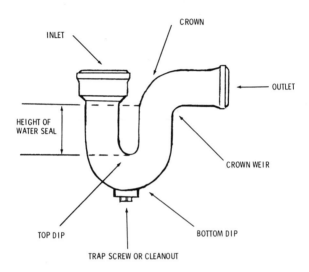

Note: Trap seal is measured from top dip to crown weir.

Note: To protect a trap water seal from evaporation in a building to be unoccupied for a period of time; pour a thin film of oil into trap.

In cold seasons water should be drained and replaced with kerosene.

Note: Trap seals may be lost by: siphonage, evaporation, cappillary attraction or wind blowing.

Note: Standard dtrap seal (2") (51 mm); seals over (2½") (64 mm) are called "deep seals."

Working Drawings
GARBAGE DISPOSAL, RESIDENTIAL

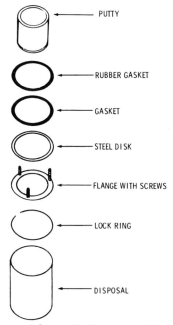

Note: Do not tighten all the way until disposal is set up and positioned for trap.

Secure disposal to waste line, tighten slip nut, or nuts do sweating last.

On two compartment sinks, disposal is roughed low, around 16" (41 cm) sink around 20" (50 cm) or 21" (53 cm).

No garbage disposal unit shall be installed upon any fresh air indirect waste system, or into any grease interceptor.

WORKING DRAWINGS
TUB VIEW IN ROUGH

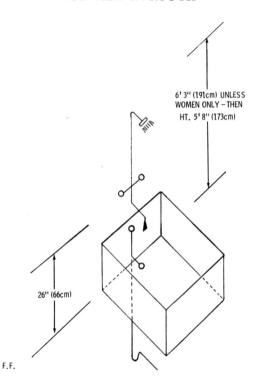

Waste 1½" (38 mm) off rough wall on center line of waste. P—Trap top should be no higher than 7" (18 cm) below floor level.

Note: Put drain piece on last.

Note: Shower rod 6'4" (193 cm) high, 27" (69 cm) off finished wall.

PRESSURE REDUCING VALVE

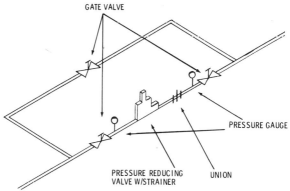

Note: When pressure exceeds 80 P.S.I. (552 kPa) a pressure regulator should be installed, especially where such pressure leads to hot water heater.

Note: Check Local Code.

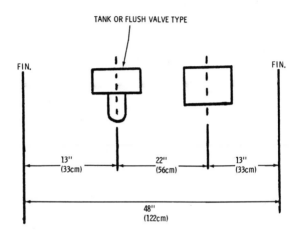

A bathroom which includes one water closet and one 20" (51 cm) lavatory can be placed in a minimum space of 48" (122 cm); finish to finish.

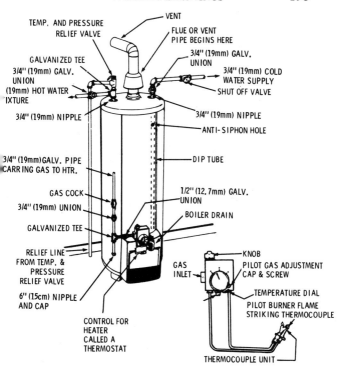

GAS HOT WATER HEATER IN AVERAGE RESIDENCE

Notes: Relief line from relief valve is generally piped to outside, 12" (305 mm) or less above ground level, elbow and nipple turned down. Nipple should not be threaded on outlet end.

The dip tube on cold water supply should terminate 8" (20 cm) above bottom of tank.

In some water heaters a special device is used to prevent corrosion. This device is called a "protector rod."

The temp.-sensitive element of the relief valve should be installed directly in the tank proper so that it is in direct contact with the hot water.

The recommended and safest position is to place relief valve in a separate tapping either at the top of the tank or within four inches from the top if tapping is located at the side.

GAS WATER HEATER

If no separate tappings are on water heater then place relief valve as shown; however use as short nipples as possible.

Never install a check valve in the water supply to a water heater, as it would confine pressure in the tank and result in an accident if the relief valve did not operate.

There is a small hole drilled in the dip tube near the top; this small hole admits air to the cold water piping to break siphonic action.

The nipple and cap at the bottom of tee where gas supply turns into heater forms a dirt and drip pocket.

PROCEDURE FOR LIGHTING HEATER

1. Turn gas cock handle on control to "off" position, and dial assembly to lowest temp. position.
2. Wait approximately 5 minutes to allow gas which may have accumulated in burner compartment to escape.
3. Turn gas cock handle on control to 'pilot" position.
4. Fully depress set button and light pilot burner.
5. Allow pilot to burn approximately 1 minute before releasing set button. If pilot does not remain lighted, repeat operation.
6. Turn gas cock handle on control to "on" position and turn dial assembly to desired position. The main burner will then ignite.

Note: Adjust pilot burner air shutter (if provided) to obtain a soft blue flame.

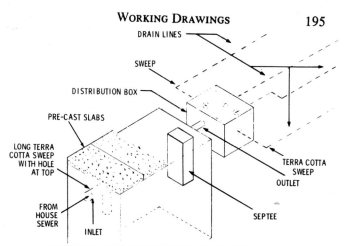

TYPICAL SEPTIC TANK INSTALLATION

Note: Information may vary according to location and type of soil.

Check Local Code.

Note: The following data is for hard compact soil:

2-bedroom house—750 gal. (2839 liter) cap. with 200 feet (61 m) of drain field

3-bedroom house—1000 gal. (3785 liter) cap. with 300 feet (91.5 m) of drain field

4-bedroom house—1000 gal. (3785 liter) cap. with 400 feet (122 m) of drain field

Note: In sandy soil smaller tanks and less drain field footage is generally the rule.

Note: Top of tank is generally 6 inches (15 cm) to 10 inches (25 cm) below ground level.

Note: Inlet invert (or bottom part of inlet) to septic tank, from top of tank, is generally 12 inches (305 mm).

Outlet Invert is generally 2 inches (51 mm) lower.

Note: If drain lines must run parallel to each other, these lines should be at least 10 feet (3.05 m) apart.

Note: Terra cotta or cement drain tile is used, measuring 4 inches (102 mm) inside and 12 inches (31 cm) long.

BEGINNING DRAIN LINES

After tank is set you begin drain field by placing one drain tile between outlet opening and distribution box.

At each opening leading to a drain line a ditch is dug 6 inches (.51 cm) below outlet opening of distribution box approximately 24 inches (61 cm) wide; wooden pegs are then driven in ground, beginning at each outlet of box and spaced every 12 to 18 inches (31 to 46 cm) apart; top of peg level with bottom of outlet openings in distribution box. Pegs should be laid level or pitched approximately 1 inch (25 mm) in 100 feet (30.5 m).

Next, the crushed rock or gravel is installed to level of pegs or 6 inches (15 cm).

Now we begin laying the drain tile, being careful to space each tile approximately ⅜" (10 mm) apart; these cracks or spaces are to be covered with tar paper.

After tile is set each drain line will then receive more crushed rock until rock reaches 1 inch (25 mm) above drain tile.

The last step is to cover entire drain line with tar paper, and cover with earth.

Note:

1000 gallon (3785 liter) septic tank contains approximately 133¾ cu. ft. (3.787 cu. m) of space. This size tank could measure

>42" (107 cm) wide
>84" (214 cm) long
>66" (168 cm) deep

750 gallon (2839 liter) septic tank contains approximately 100¼ cu. ft. (2.838 cu. m) of space. This size tank could measure

>35" (89 cm) wide
>72" (183 cm) long
>68¾" (175 cm) deep

Note: Measurements are inside dimensions.

FORMULAS

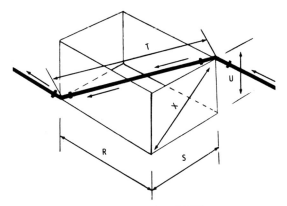

ROLLING OFFSETS

Formula using 45° Fitting
$X = \sqrt{S^2 + U^2}$
$T = \text{"X"} \times 1.41$
$R = X$

Formula using 60° Fittings
$X = \sqrt{S^2 + U^2}$
$T = \text{"X"} \times 1.15$
$R = \text{"X"} \times .58$

Formula using 22½° Fittings
$X = \sqrt{S^2 + U^2}$
$T = \text{"X"} \times 2.61$
$R = \text{"X"} \times 2.41$

"EXAMPLE"

Using 45° Fittings assume "U" is 25 and "S" is 30

"U" must be squared

STEP 1

$$\begin{array}{r} 25 \\ \times\ 25 \\ \hline 125 \\ 50\ \ \\ \hline 625 = \text{"U" squared} \end{array}$$

STEP 2 "S" must be squared

$$\begin{array}{r} 30 \\ \times\ 30 \\ \hline 900 = \text{"S" squared} \end{array}$$

STEP 3 Add U = 625
S = 900
Total 1525
Find Square Root of 1525

$$\begin{array}{r} 3 \\ \text{X}\ 2 \\ \hline 6 \end{array}$$

$$\begin{array}{r} 60 \\ 9 \\ \hline 69 \\ 9 \\ \hline 621 \end{array}$$

R4
3 9 = Sq. Root
√1525
9
625
621
4

Proved
39
× 39
351
117
1521
± 4 = R
1525

STEP 4

X = 39

"T" = Travel or Pipe to be cut

55 -C-C

$$\begin{array}{r} 1.41 \\ \times\ 39 \\ \hline 1269 \\ 423\ \ \\ \hline 54.99 \end{array}$$

Answer: 55 -C-C

FORMULAS

SQUARE ROOT—REVIEWED

STEP 1

Pointing off number to be worked in preparation to solution is first and most important step.

Whether decimal or whole number, always point off in two's starting from right to left. When decimal occurs and numbers are odd a zero (0) must be added thusely

$\sqrt{25.325}$, should be $\sqrt{25.3250}$.

Examples: "Pointing Off."

$$\sqrt{2'59} \qquad \sqrt{25'32} \qquad \sqrt{25.'32'50}$$

STEP 2

Beginning with 1st number or set of numbers on left we begin our problem; in the three examples above the numbers would be (2) (25) (25) respectively.

We must find nearest square (or number multiplied by itself) to this number but not to total higher than number.

In examples, square's would be: (1), (5), (5) respectively.

EXAMPLE

STEP 3

```
    5                             5. 0 3      R 241
  × 5         1000             √ 25.'32'50
  ---         + 3                25
   25         ----                ------
              1003                3250
              × 3                 3009
              ----                ------
              3009                 241 R
```

The nearest square as indicated above, was (5) is now multiplied by (2) and brought down to continue problem $\begin{array}{r} 5 \\ \times\ 2 \\ \hline 10 \end{array}$	**PROBLEM PROVED:** 25.3250 5.03 $\times$ 5.03 1509 25150 25.3009 + 241 = R 25.3250

STEP 4
After (5) is doubled and brought down a zero (0) is then added.

STEP 5
The next two numbers are brought down (32), since 100 can not be divided into 32 a zero (0) is placed above in answer. A (0) added to 100, and next two numbers brought down.

STEP 6
We now will determine number of time (1000) will go into (3250); the number being (3); we now add this number to 1000; resulting in 1003; we then multiply by 3 and enter this figure below 3250. The number (3) is placed above.

(241) is our remainder unless we wish to carry it out further; if so two zeros (00) must be added then brought down, added to 241, and problem continued.

NOTE:
Decimals in a problem are pointed off in two's but equal (1) in answer. In our example .3250 is four decimal places but pointing off in two's it will read two in answer.

FINDING DIAGONAL OF A SQUARE

NOTE:
The diagonal of a square equals the square root of twice the area.

EXAMPLE:
Find the diagonal of a square when area is 8100 square inches.

16,200 sq. ins. = twice the area

```
   8,100
+  8,100
  ------
  16,200 sq. ins.
       = twice the area
```

Note: Figures used in example could also be centimeters or meters, etc.

90"
90"
Diagonal
127¼"

```
     20    │  1 2 7. 2
   +  2    │ √1'62'00.00
   ----    │  1
     22    │  --------
           │  62
    240    │  44
   +  7    │  --------
   ----    │  1800
    247    │  1729
           │  --------
   2540    │  7100
   +  2    │  5084
   ----    │  --------
   2542    │  2016ᴿ
```

Ans.: 127¼"
Rounded

Note:
Review Step 3 on pages 199 and 200 under Square Root.
Reviewed.

FORMULAS

RIGHT TRIANGLE

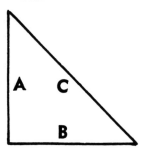

C = Hypotenuse
A = Altitude
B = Base

Formula — $C^2 = A^2 + B^2$

NOTE:
The square of the hypotenuse equals the sum of the squares of the other two sides.

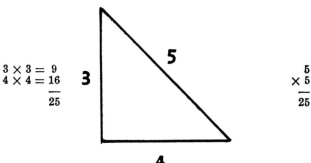

NOTE:
To square your work while fabricating, or form a square use 3 — 4 — 5 method. You may use 3 ft., 4 ft., 5 ft., or you may use 3", 4", 5"; to increase sides of square merely double or triple each number;
Examples: (6 — 8 — 10) 9 — 12 — 15) and so on.

FORMULAS

SUBJECT:

Running pipe or tubing in parallel runs using offsets and maintaining uniform spread throughout run."

Table for Figure (1)

22½° Ells or 1/16 Bends	45° Ells, or ⅛ Bends	60° Ells, or 1/6 Bends
$T = U \times 2.61$	$T = U \times 1.41$	$T = U \times 1.15$
or	or	or
$T = R \times 1.08$	$T = R \times 1.41$	$T = R \times 2$
$U = R \times 0.41$	$U = R$	$U = R \times 1.73$
$R = U \times 2.41$	$R = U$	$R = U \times 0.58$
$X = H \times 0.20$	$X = H \times 0.41$	$X = H \times 0.58$

Formulas for Figure (2)

$X = H \times 0.41$

$B = A + (H \times 0.41 \times 2)$ or
$A = B - (H \times 0.41 \times 2)$

$C = B + (H \times 0.41 \times 2)$ or
$B = C - (H \times 0.41 \times 2)$

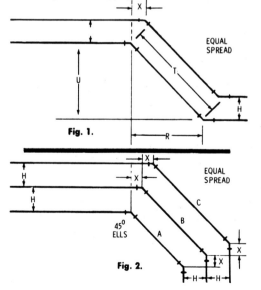

Fig. 1.

Fig. 2.

204 FORMULAS

Figure Below Illustrates:
  45° Y and 1/6 Bend in Offset

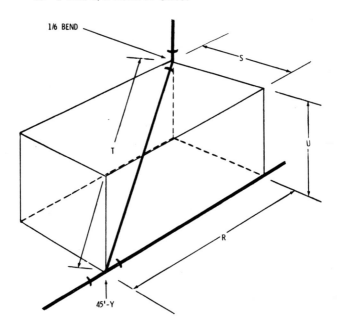

Formula for Above:

$S = U$
$R = S \times 1.41$
$T = S \times 2$, or $U \times 2$, or
  $R \times 1.41$
U = Vertical Rise
S = Horizontal Spread
R = Advance or Setback
T = Travel or Pipe to be cut
"T" will be C-C-Measurement

FORMULAS

DETERMINING CAPACITY IN GALLONS OF A TANK

Example: Tank is 4′ 0″ in Diameter and 10′ 0″ long—
(A) Find the area of the circle;
    Diameter squared × .7854 :
Thus, 4′ × 4′ × .7854:
Thus, 4′ × 4′ × .7854 = 12.566 square feet.

(B) Find the cubical contents:
Area of a circle × Length:
12.566 sq. Ft. × 10′ = 125.66 **cu. feet.**

(C) Find the number of gallons:
Multiply the cubical contents by 7.48 (the number of gallons in one cubic ft.):
125.66 cu. ft. × 7.48 gals. = 939.936, or "940 gals."
Answer to Example.

DETERMINING CAPACITY IN LITERS OF A TANK

Example: Tank is 1.22 meters in diameter and 3.05 meters long—

(A) Find the area of the circle;
    Diameter squared × .7854:
Thus, 1.22 × 1.22 × .7854:
Thus, 1.22 × 1.22 × .7854 = 1.169 sq. meters

(B) Find the cubical contents:
Area of a circle × Length:
1.169 sq. meters × 3.05 meters = 3.565 cu. meters

(C) Find the number of liters:
Multiply the cubical contents by 1000 (the number of liters in one cubic meter):
3.565 cu. meters × 1000 = 3565 liters.

FORMULAS

Area of Triangle — $A = \frac{1}{2} BH$

Area of Circle — $A = \pi R^2$
or $.7854 D^2$

Note: $\pi = 3.1416$

To find area of circle when circumference is known:
$$A = \frac{C^2}{12.57}$$

If Circle = 45
$$A = \frac{45 \times 45}{12.57}$$

$A = 161.09$ Answer

To find the "Head" when pressure is given divide pressure by .433 —or multiply the pressure by 2.309.

Formula — $PSI = H \times .433$
$H = PSI \times 2.309$

ROLLING OFFSET
FURTHER SIMPLIFIED

Fraction — Change to decimal when multiplying

Fraction	Decimal
1/16"	.06
1/8 "	.13
3/16"	.19
1/4 "	.25
5/16"	.31
3/8 "	.38
7/16"	.44
1/2 "	.50
9/16"	.56
5/8 "	.63
11/16"	.69
3/4 "	.75
13/16"	.81
7/8 "	.88
15/16"	.94

Formulas

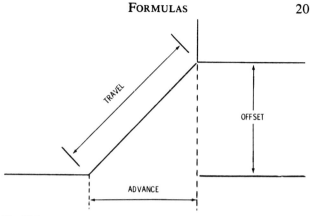

TABLE OF CONSTANTS
FOR CALCULATING OFFSET MEASUREMENTS

Degree of Fitting	Known Factor	Constant	
5⅝°	Offset ×	10.20	= Travel
5⅝°	Offset ×	10.152	= Advance
11¼°	Offset ×	5.126	= Travel
11¼°	Offset ×	5.027	= Advance
22½°	Offset ×	2.613	= Travel
22½°	Offset ×	2.414	= Advance
30°	Offset ×	2.000	= Travel
30°	Offset ×	1.732	= Advance
45°	Offset ×	1.414	= Travel
45°	Offset ×	1.00	= Advance
60°	Offset ×	1.155	= Travel
60°	Offset ×	0.577	= Advance
67½°	Offset ×	1.083	= Travel
67½°	Offset ×	0.414	= Advance
72°	Offset ×	1.052	= Travel
72°	Offset ×	0.325	= Advance

NOTE: When above constants are used to calculate offset measurement, given offset and solution are expressed in center to center measurements.

FORMULAS

CONVERTING DECIMAL PARTS OF A FOOT TO INCHES:

1″ = 0.083′
2″ = 0.1666′
3″ = 0.25′
4″ = 0.333′
5″ = 0.4167′
6″ = 0.50′
7″ = 0.5833′
8″ = 0.6667′
9″ = 0.75′
10″ = 0.8333′
11″ = 0.9333′
12″ = 1.00′

To convert decimal feet to inches, multiply by 12″: then you may change decimal inches to inches and fractions. Example:

```
      .6875'
     × 12"
    -------
    1 3750
    6 875
    -------
    8.2500"  or  8¼"
```

METRIC INFORMATION

English "Parts of an Inch"	Metric Equiv. "Millimeters"	
1/32″	.79375	(.80)
1/16″	1.5875	(1.6)
1/8″	3.175	(3.2)
3/16″	4.7625	(4.8)
1/4″	6.35	(6.4)
5/16″	7.9375	(7.9)
3/8″	9.525	(9.5)
7/16″	11.1125	(11.1)
1/2″	12.7	(12.7)
9/16″	14.2875	(14.3)
5/8″	15.875	(15.9)
11/16″	17.4625	(17.5)
3/4″	19.05	(19.1)
13/16″	20.6375	(20.6)
7/8″	22.225	(22.2)
15/16″	23.8175	(23.8)
1″	25.4	(25.4)

FORMULAS

Fahrenheit		Celsius
212°	Temp. of Boiling Water	100°
176°		80°
140°		60°
122°		50°
104°		40°
98.6°	Temp. of Human Body	37°
95°		35°
86°		30°
77°		25°
68°		20°
50°		10°
32°	Temp. Melting Ice	0°
− 4°		− 20°
− 40°	Temp. Equal	− 40°

The following formula may be used for converting temperatures given on one scale to that of the other.

$F = 1.8C + 32$
    or $1.8 \times C + 32$

$C = F - 32 \div 1.8$
    or F minus 32 divided by 1.8

Symbols for Plumbing Fixtures
"SYMBOLS FOR PLUMBING FIXTURES"

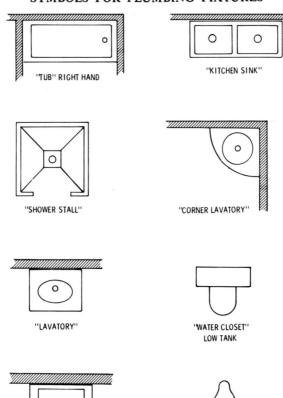

Symbols for Plumbing Fixtures

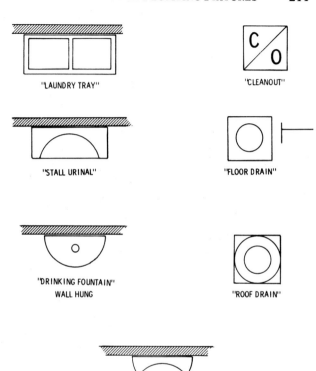

OXYACETYLENE BASIC SAFETY MEASURES

I wish to thank the Compressed Gas Association, Inc., New York, New York, for providing the following information on safe practices to follow when oxyacetylene welding or cutting.

1. Always blow out cylinder valves before attaching regulators. Dust can cause combustion resulting in an explosion.

2. Stand to side of regulator when opening cylinder valve. The weakest point of every regulator is either front or back. Regulator could blow out, an explosion could occur.

3. Always release adjusting screw on regulator before opening oxygen cylinder valve.

When the adjusting screw is released, the seat of the regulator is in contact with the nozzle and with sufficient pressure to hold the 2200 PSI (15168 kPa). So the oxygen released travels only a short distance.

If regulator were open we release the high pressure through the seat nozzle and we have expansion going into the regulator, then restriction into the nozzle, thus generating a lot of heat that could set off dust or oil.

Oxyacetylene Basic Safety Measures 213

4. Always open cylinder valve slowly.

By opening the valve slowly, the heat made from the travel is very small. The main reason is to reduce shock. If valve is opened fast, the pressure exerted from the shock hitting the seat surface exceeds that of the pressure contained in the cylinder.

"Open slowly and save repair bills plus eliminating the explosion danger."

5. A good practice is to light fuel gas before opening oxygen valve on torch.

Light large tip and show soot, then no soot, then flame leaving tip. The burning rate should be set with the acetylene valve only. If you do open the oxygen valve first, you pop the large tip.

6. Never use oil on regulators, torches etc.

Oxygen and oil create an explosion.

Note: In oxygen cylinders there is as much as 2200 PSI (15168 kPa) pressure. When the pressure is released from the cylinder through the regulator, the speed at which the oxygen travels exceeds the speed of sound and this generates heat and friction. (The smallest amount of oil, just the oil from your skin, will ignite and blow up the regulator.)

7. Do not store cylinders near flammable material, especially oil, grease, or any other readily combustible substance.

8. Acetylene cylinders should be stored in a dry and well ventilated location.

9. Acetylene cylinders should not be stored in close proximity to oxygen cylinders.

10. Never tamper with safety devices in valves or cylinders. Keep sparks and flame away from acetylene cylinders and under no circumstances allow a torch flame to come in contact with safety devices. Should the valve outlet of an acetylene cylinder become clogged by ice, thaw wth warm—(not boiling)—water.

11. Acetylene should never be used at a pressure exceeding 15 PSI (103 kPa) gauge.

12. The wrench used for opening the cylinder valve should always be kept on the valve spindle when the cylinder is in use.

13. Finally! Points of suspected leakage should be tested by covering them with soapy water. "NEVER test for leaks with an open flame."

GENERAL INFORMATION

In this section I have placed information that I considered useful, informative, and most generally applied to "plumbing of today."

"Plumbing" Defined

It's the art and science of creating and maintaining sanitary conditions in buildings used by humans. It's also the art and science of installing, repairing, or servicing in these same buildings, a plumbing system which includes the pipes, fixtures, and appurtenances necessary for bringing in the water supply, and removing liquid and water carried waste.

PURPOSE OF VENTING IN PLUMBING DRAINAGE SYSTEM

To provide equal pressure in plumbing system:

Venting prevents pressures from building up causing retarded flow.

Protects trap seals, carries off foul air, gases, vapors—which would form corrosive acids harmful to piping.

A "Plumbing System" is designed for not more than 1" (2.5 cm) pressure at the fixture trap. Greater pressures may disturb the trap seals. (1" (2.5 cm) of pressure) is a pressure equivalent to that of a 1" measuring the same water column.

A "Loop Vent" is the same as a "Circuit Vent" except that it loops back and connects with a "Stack Vent" instead of a "Vent Stack."

A waste stack terminates at the highest connection from a fixture. From this point to its terminal above the roof it is known as a stack vent.

SAFETY REMINDERS

Many persons have been blinded by getting lime into the eyes either as a powder or as mortar. If you should get lime into your eyes, wash them clean at once with water. Go to a good doctor for further attention. Lime touching skin: wash with water, rinse with vinegar, then coat with vaseline in which ordinary kitchen soda bicarbonate has been mixed.

When using electric power tools with abrasive cutting wheels be sure tools are properly grounded, eyes and hands properly protected.

When using a ladder, the ladder base should be placed one-fourth of the ladder length away from the structure against which the ladder is leaning.

One of the best agents to use when fighting fires that occur in or around electrical equipment is a dry powder Co_2 Extinguisher. Which is actually composed of 99% baking soda and 1% drying agent.

"SAFETY" — A good item to have on any job requiring the use of an open flame, such as a prestolite torch or melting pot tank, or welding equipment is an approved type Co_2 or Dry Powder extinguisher.

Note: A carbon tetrochloride extinguisher at one time was considered as the best agent to use but it was found to release a deadly gas. Dangerous if used in a confined place.

GENERAL INFORMATION
KNOTS COMMONLY USED

Fig. 1. The Bowline

Fig. 2. The Square Knot

Fig. 3. The Bowline on a Bight

GENERAL INFORMATION
KNOTS COMMONLY USED

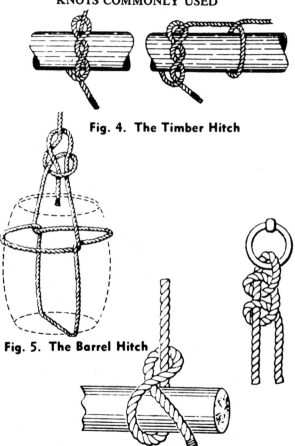

Fig. 4. The Timber Hitch

Fig. 5. The Barrel Hitch

Fig. 6. The Half Hitch & Two Half Hitches

GENERAL INFORMATION
TYPICAL HOISTING SIGNALS

1. **Hoist Load**

2. **Lower Load**

3. **Stop**

4. **Swing or House**

5. **Boom Up**

6. **Boom Down**

7. **Travel Forward**

8. **Travel Backward**

9. **Move Slowly**

10. **Emergency Stop**

GENERAL INFORMATION

COMPRESSION TANK

In heating systems this tank plays an important part in the economical operation of the system. When water in the system is heated it expands and if no tank were installed, the expanding water would be forced out through the relief valve.

When system cooled additional water would be drawn in to replace the water lost by expansion.

Extra fuel is used to heat this cold water; constant adding of water brings in foreign matter, such as sediment or lime. This results in scaling of boiler with an ever increasing amount of fuel required for heating."

NOTE:

If a 30 gal. (113.56 liter) hot water boiler is insulated with a tank jacket 30% of total amount of gas usually burned can be saved.

If hot water pipes are insulated, the heat loss from pipes is reduced by up to 80%.

There should be a minimum of 6" (15 cm) between uninsulated water heater and any unprotected wood.

NOTE: Water in a heating system, heated from 32°F. to 212°F. (0°C. to 100°C.) will expand approximately 1/23 of its original volume.

The dip tube on cold water supply should terminate 8" (20 cm) above bottom of tank.

In some water heaters a special device is used to prevent corrosion, this device is called a "Protector Rod."

General Information

To determine or form an angle by use of rule is as follows:

22½° or 1/16th bend — take tip of rule and touch (23¾") straighten out rule at second joint for angle.

11¼° or 1/32nd bend — take tip of rule and touch 23-15/16". Follow as above.

30° Angle — touch tip of rule to 15⅜". Follow as above.

45° — or ⅛th bend — touch 23" with tip. Follow as above.

60° or 1/6th bend — touch 22¼" with tip. Follow as above.

72° Angle — touch tip of rule to 21⅝". Follow as above.

90° Angle — touch tip of rule to 20¼". Follow as above.

Subject—Garbage disposals. Cold water must be used as it congeals the grease particles. Mixing them with food particles together being flushed down drain.

Hot water — liquifies grease and if constantly used a stoppage would eventually occur due to accumulated coatings of grease.

Size of pipe wrench is measured from inside top of movable jaw to end of handle with wrench fully opened.

Use light machine oil when oiling a rule.

Hacksaw blade manuf. recommend that a blade with 24 teeth per inch (10 teeth per cm) be used for cutting angle iron or pipe; when same blade is desired to cut hanger rod (18 teeth per in. or 7 teeth per cm) is satisfactory.

Light gauge band iron and thin wall tubing (32 teeth per in. or 13 teeth per cm) is best.

GENERAL INFORMATION 223
MACHINE SCREW "BOLT" INFORMATION "N.C."

Diameter in Inches or millimeters

Size	Decimal	To Nearest 1/64" (½ mm)
1	.0730	5/64" (2 mm)
2	.0860	5/64" (2 mm)
3	.0990	7/64" (2½ mm)
4	.1120	7/64" (3 mm)
5	.1250	1/8" (3 mm)
6	.1380	9/64" (3½ mm)
8	.1640	11/64" (4 mm)
10	.1900	13/64" (5 mm)
12	.2160	13/64" (5½ mm)
1/4" (6.35 mm)	.2500	1/4" (6 mm)

Note: Above No. 12, machine screw sizes are designated by actual diameter.

STANDARD WOOD SCREW INFORMATION

Diameter in Inches or millimeters

Size Number	Decimal	Approx. Fraction or ½ mm
0	.064	1/16" (1½ mm)
1-2	.077, .090	3/32" (2 mm)
3-4-5	.103, .116, .129	1/8" (2½ to 3 mm)
6-7-8	.142, .155, .168	5/32" (3½ to 4 mm)
9-10-11-12	.181, .194, .207, .220	3/16" (4½ to 5½ mm)
14-16	.246, .272	1/4" (6 to 7 mm)
18-20	.298, .324	5/16" (7½ to 8 mm)
24	.376	3/8" (9½ mm)

"Flathead Screws" are measured by over-all length; "Round-Head Screws" from base of head to end.

ABBREVIATIONS

A.G.A. — American Gas Association
A.S.A. — American Standards Association
A.S.H.V.E. — American Society of Heating & Ventilation
F. & D. — Faced and Drilled
I.B.B.M. — Iron Body Bronze or Brass Mounted
M.S.S. — Manufacturer Standardization Society of Valve & Fittings Industry
N.P.S. — Nominal Pipe Size
O.S. & Y. — Outside Screw & Yoke
L.I.A. — Lead Industries Association
R.N.P.T. — Right Hand Nat. Pipe Thread
N.D.T.S. — Not Drawn to Scale
A.S.T.M. — American Society for Testing Material
C.A.B.R.A. — Copper & Brass Research Association
C.I.S.P.I. — Cast Iron Soil Pipe Institute
B.M. — Bench Mark

AMERICAN STANDARD THREADS

Pipe Size	Threads Per Inch	Threads Per Centimeter
⅛" or .3175 cm	27	10½
¼" or .635 cm	18	7
⅜" or .9525 cm	18	7
½" or 1.27 cm	14	5½
¾" or 1.905 cm	14	5½
1" or 2.54 cm	11½	4½
1¼" or 3.175 cm	11½	4½
1½" or 3.81 cm	11½	4½
2" or 5.08 cm	11½	4½
2½" thru 4" or 6.35 thru 10.16 cm	8	3⅛

TABLE OF SQUARES

$11^2 = 121$ $19^2 = 361$
$12^2 = 144$ $20^2 = 400$
$13^2 = 169$ $21^2 = 441$
$14^2 = 196$ $22^2 = 484$
$15^2 = 225$ $23^2 = 529$
$16^2 = 256$ $24^2 = 576$
$17^2 = 289$ $25^2 = 625$
$18^2 = 324$ $26^2 = 676$

GENERAL INFORMATION
"HANGER ROD SIZES"

Iron Pipe Size	Rod Size
1/8"-1/2" (3-13 mm)	1/4" (6 mm)
3/4"-2" (19-51 mm)	3/8" (10 mm)
2½" and 3" (64 and 76 mm)	½" (13 mm)
4" and 5" (10 and 13 cm)	5/8" (16 mm)
6" (15 cm)	3/4" (19 mm)
8", 10", 12" (20, 25, 30 cm)	7/8" (22 mm)
14" and 16" (36 and 41 cm)	1" (25 mm)

Hanger rod is threaded with (N.C.) national coarse bolt dies.

A threading die marked 5/8"-11-NC— would be used for 5/8" diameter rod, the thread classified as "National Coarse" with 11 threads per inch.

A "Vacuum Relief Valve" would be installed on a copper tank to prevent collapse in the event of a vacuum occurence.

Average rate of rainfall is 4" or (10 cm) per hour.

75' (23 m) maximum spacing for rainwater leaders.
150 sq. ft. (14 sq. meters) of roof area to 1 sq. inch (6.5 sq. cm) of leader area.
The instrument for measuring relative humidity is called a Hygrometer.

Transfer of heat occurs in three ways:
    (1) convection
    (2) radiation
    (3) conduction

"Bread" can be packed in a water line to hold back water long enough to solder a joint where water continues to trickle, despite valve or valves having been shut off.

MELTING POINTS

Lead melts at 622°F. (328°C.)
Tin melts at 449°F. (231.7°C.)
50-50 solder begins to melt at 362°F. (183.3°C.)
Zinc melts at 790°F. (421°C.)
Pure Iron melts at 2,730°F. (1,499°C.)
Steel melts from 2,400 to 2,700°F., (1,315.5°C. t 1,482.2°C.)

Cylinders are charged with oxygen at a pressure c 2000 PSI (13789 kPa) at 70°F. (21°C.).

The temperature of oxygen-acetylene flame estimate to be over 6000°F (3316°C.).

"Gasket Material" to be suitable in a flange unio should be made of:

"Cold Water Piping"—sheet rubber or asbestos shee packing.

"Hot Water Lines"—rubber or asbestos composition

"Gas Piping"—leather or asbestos composition.

"Oil Lines"—metallic or, where permitted, asbesto composition.

"Gasoline Conduction"—metallic.

By applying graphite to one side of a gasket removal a a later date is made much easier.

General Information

"Datum" is an established level or elevation from which vertical measurements are taken; A "Bench Mark" on which all other elevations are based.

All buildings have a "base elevation" from which all other elevations and grades are determined; some plans use 100.0', others use 0.00'

"Base Level" ---------------------------------B.M.

Example:

Basement floor level 91.5', using 100.0' as the "base level" would indicate basement floor level 8' 6" below 1st floor level.

B.M. means "bench mark"

"Bench marks" permit the plumber to locate the elevations applying to his work. A 96' or a 104' bench mark would indicate 4' below or 4' above "finish floor." Examples of "bench marks"

$$\text{B.M.} \quad \overset{\wedge}{\rule{1cm}{0.4pt}} \text{ F.F.} + 4.0'$$

$$\overset{\wedge}{\rule{1cm}{0.4pt}} \text{ F.F.} - 2.0'$$

A Sectional Elevation Drawing" would provide the plumber with information as to "Width and Height" of a specific portion of the structure.

Elevation measurements on piping plans are called: "Invert Elevations."

General Information

1 lb. of air pressure elevates water approx. 2.31 ft. under atmos. cond. of 14.72 P.S.I.

2.3 Ft. of water = 1 P.S.I.

1 ft. of water = .434 P.S.I.

1,728 cu. ins. = 1 cu. ft.

231 cu. inches in one U. S. gallon.

One cu. foot of water at 39° F. weighs 62.48 lbs.

One U. S. gallon of cold water weighs 8.33 lbs.

One cu. ft. of water contains 7.48 gals.

"SI"

1 kPa of air pressure elevates water approx. 10.2 cm under atmospheric conditions of 101 kPa.

10.2 cm of water = 1 kPa.

51 cm of water = 5 kilopascals.

1 meter of water = 9.8 kPa.

10,000 sq. centimeters = 1 sq. meter.

1 cubic meter = 1,000,000 cu. cm or 1000 cu. decimeters.

1 liter of cold water at 4°C. weighs 1 kilogram

General Information

One cubic inch of mercury weighs .49 lbs.

Therefore 10″ column of mercury would be 10 × .49 or 4.9 PSI. Generally speaking 2″ of mercury = 1 lb. pressure.

One cu. ft. of air weighs 1.2 oz. or .075 lbs.

Atmospheric pressure of 14.7 PSI will balance or support a column of mercury 29.92″ high.

Absolute zero is 459.6° below zero.

The absolute minimum fall or grade for foundation or subsoil drainage lines is 1″ (2.5 cm) in 20′ (6.1 m).

"SI"

1 centimeter column of mercury at 0°C. = 1.3332239. kPa pressure.

Therefore, a 24 cm column of mercury would be 24 × 1.3332239 or rounded 32 kPa. Generally speaking, 6 centimeters (cm) of mercury = 8 kilopascals (kPa).

1 cubic meter of air weighs 1.214 kilograms.

Atmospheric pressure of 101.3 kPa will balance or support a column of mercury 76 cm high.

Absolute zero is −273.16°C.

GENERAL INFORMATION

An "air gap" of 1" (2.5 cm) to 2" (5 cm) between flood level of a fixture and the water supply opening is considered safe.

A "vacuum breaker" should be at least 6" (15 cm) above flood level rim or 6" (15 cm) above top of unit.

"BOILING POINTS OF WATER AT VARIOUS PRESSURES ABOVE ATMOSPHERIC"

Atmospheric, or "O" Gauge Pressure: Boiling Point 212°F. (100°C.)

"Gauge Pressure" PSI or kPa	"Boiling Point"
1—6.89	216 Degrees F. (102.2°C.)
4—27.58	225 Degrees F. (107.2°C.)
15—103.43	250 Degrees F. (121.1°C.)
25—172.36	267 Degrees F. (130.5°C.)
30—206.84	274 Degrees F. (134.4°C.)
45—310.26	293 Degrees F. (145°C.)
50—344.73	297 Degrees F. (147.2°C.)
65—448.13	312 Degrees F. (155.5°C.)
75—517.1	320 Degrees F. (160°C.)
90—620.52	335 Degrees F. (168.3°C.)
100—689.47	338 Degrees F. (170°C.)
125—861.83	353 Degrees F. (178.3°C.)
150—1034.2	336 Degrees F. (185.5°C.)

Gases found in sewer air: carbon monoxide, methane, hydrogen sulphide, carbon dioxide, gasoline, ammonia, sulphur dioxide, illuminating gas.

Absolute pressure is gauge pressure plus atmospheric.

Wrench size for flange bolt is: Bolt size × 2 + ⅛ inch (× 2 + 3 mm)

General Information

A "Cross-Connection" is any physical connection or arrangement of piping which provides a connection between a safe water supply system and a separate system or source which is unsafe or of questionable safety, and which under certain conditions permits a flow of water between the safe and unsafe systems or sources.

"Air Chambers" to be effective, should be located as close as possible to the points at which water hammer will occur.

"Water Hammer" becomes much greater at 100 PSI (689.5 kPa) than at 50 PSI (345 kPa). Two types of manufactured devices used to reduce or eliminate water hammer are "Shock Absorbers" or "Water Hammer Arrestors" to cushion the shock, and "Pressure-Reducing Valves" to lower the operating pressure.

"Globe Valves" have a machined seat and a composition disc and usually shut off tight, while "Gate Valves" may leak slightly when closed, particularly if frequently operated, due to wear between the brass gates and the faces against which they operate .Globe valves create more flow resistance than gate valves.

"Domestic Hot Water" from 140°F. to 160°F. (60°C. to 71°C.) is considered acceptable maximum temp. for domestic hot water. Use of automatic laundry and dishwashing machines makes 160°F. (71°C.) preferable. Temperatures above 160° F. (71°C.) are not recommended. They cause increased corrosion; increased deposit of lime; waste of fuel; more rapid heat loss by radiation; danger of scalding and other accidents.

General Information

The standard length of asbestos cement water main pressure pipe is 13 ft. (approximately 4 meters).

The lateral lines in a subsurface disposal field for a septic tank should be installed at a grade of 3″ per 100′ (76 mm per 30.5 meters).

The installation of a water softener in a residential piping system causes a "fairly high pressure loss".

A friction clamp for use with brass pipe in regular pipe vise is made by cutting a pipe coupling in half lengthwise—then lining with "sheet lead."

When a line is neither horizontal or vertical it is said to be slanting or diagonal.

Brass pipe expands about 1¼″ (32 mm) per 100 ft. (30.5 meters) for 100°F. (38°C.) rise in temp.

Steel pipe expands about ¾″ (19 mm) per 100 ft. (30.5 meters) for 100°F. (38°C.) rise in temp.

Storage tanks up to 82 gal. (310.4 liter) Cap. are tapped for 1″ (25 mm) Conn.; tanks over that size are tapped a min. of 1¼″ (32 mm) generally.

Brass Fittings contain 85% Copper, 5% zinc, 5% tin, 5% lead.

NOTE:
  18" (46 cm) of 4" (10 cm) pipe holds one gal. (3.785 liter) of water.

The max. ht. of horses supporting scaffold platforms is 16 ft. (4.88 meters).

In trenches in hard compact ground vertical braces should be located every 8 ft. (2.44 meters) and horizontal stringers every 4 feet (1.22 meters).

A figure in isometric position lies with one corner directly in front of you. The back corner tilted to a 30° angle.

When water solidifies it becomes lighter.

The amount of heat required to change ice to liquid water is 144 B.T.U.'s per lb. (335 joules per kilogram).

Approx. hts. above F.F. or Flood Level rims for plumbing fixtures are: Sink—36" (91 cm), Built-in Bathtub—16" (41 cm), water closet—15" (38 cm), lavatory—31" (79 cm), wash or laundry tray—34" (86 cm).

Pipe above 12" (305 mm) diameter is generally classified by its outside diameter. Thus, 14" (356 mm) pipe would be it its O.D.

The "Invert" of a "Sewer Line" or any pipe line is the inside wall flow line at bottom of pipe.

GENERAL INFORMATION 234

2% grade is slightly less than ¼" per ft. Example: sewer 220' long

$$\begin{array}{r} 220' \\ .02 \\ \hline 4.40 \end{array}$$ ft. or approx. 52¾" total fall

Example in metric—sewer 37 meters

$$\begin{array}{r} 37 \\ \times .02 \\ \hline .74 \end{array}$$ meters or 740 mm = Total fall in 37 meters on 2% grade.

When a soldering iron is overheated, the bright areas will show bluish tarnish, the tinning on the bit will be dark, dull, and powdery in appearance.

When copper and steel come in contact with each other, especially when dampness is present, a chemical action called (Electrolysis) is created.

Suction pumps, barometers and syphons depend upon the natural pressure of atmosphere in order to function.

A "Vacuum" is a space entirely devoid of matter. A "partial vacuum" is a space where there exists an air pressure less than atmospheric.

In colder climates, the closer the "vent terminal" is to the roof, the less chance of (frost closure).

The "Building Sewer" extends from the main sewer or other disposal terminal to the "Building Drain" at a distance of approx. 5' (152 cm) from the foundation wall.

General Information

The "Building Drain" is the lowest horizontal piping inside the building, it connects with the building sewer.

"Lead Pans" for (Shower Stalls) installed on new concrete floors should be given a heavy coating of (Asphaltum) both inside and out. The Asphaltum protects the lead from "corrosion" during the curing period of the concrete, due to chemical reaction created between concrete, lead and the water seeping through and contacting both.

"Convection" is the method used for transferring heat in a (gravity domestic hot water circulation system.)

"Convection," or circulating currents, are produced due to the difference in weight of water at different temperatures.

"Condensation" is formed on (cold water pipe lines) when warm "humid air" comes in contact with the cold surfaces causing these pipe lines to give up some of their moisture.

A "Corporation stop" is located at the tap in the city water main. The Connection between the "City Water Main" and the building is called the (service pipe).

"Public Sewer Manholes" can be used to verify (main sewer elevations) and direction of flow.

A 1-inch (25 mm) "water pipe" is equal to four 1/2 inch (13 mm) pipes.

"Water Supply Pipes" extending vertically one or more stories are called (risers).
"Soil and vent pipes" extending vertically are called (Stacks).

"Primary Treatment" in a (Sewage Treatment Plant) removes floating and settleable solids.
"Secondary Treatment" removes dissolved solids.

PLASTIC PIPE AND FITTINGS

I wish to thank the Plastics Pipe Institute, 355 Lexington Avenue, New York, New York 10017, for making the following section on Plastics possible. I am very proud to include this section, it is my opinion that the information contained herein will prove to be very beneficial and informative to the Plumbing Industry, and all other individuals interested enough to follow directions.

Plastic DWV Piping has been approved (at this writing) by local codes within 32 states, including the Building Officials Conference of America, Southern Building Code Congress, International Association of Plumbing and Mechanical Officials, and FHA.

PVC—Type 1 Polyvinyl Chloride is strong, rigid and economical. It resists a wide range of acids and bases, but may be damaged by some solvents and chlorinated hydrocarbons. Maximum service temperature is 140°F. (60°C.). PVC is better suited to pressure piping.

ABS—usage is almost double, as compared with PVC in DWV piping systems; however, it is limited to 160°F. (71.1°C.) water temperatures at lower pressures considered adequate for DWV use.

CPVC—meets National Standards for piping 180°F. (82.2°C.) water at pressures of up to 100 PSI (689 kPa), and can withstand 200°F. (93.3°C.) water temp. for limited periods. Chlorinated Polyvinyl Chloride is similar to PVC in strength and overall chemical resistance.

Polypropylene—is a material very lightweight and suitable for lower pressure applications up to 180°F. (82.2°C.). It is used widely for industrial and laboratory drainage of acids, bases and many solvents.

Kem-Temp (PVDF) or Polyvinylidene—Fluoride is a strong, tough and abrasive resistant fluorocarbon material. It has excellent chemical resistance to most acids, bases and organic solvents and is ideally suited for handling wet or dry chloride, bromine and other halogens. It can be used in temperatures of up to 280°F. (138°C.).

FRP EPOXY—is a fiberglass reinforced thermoset plastic with high strength and good chemical resistance up to 220°F. (104.4°C.).

EXPANSION IN PLASTIC PIPING

PVC-Type 1—100′ or 30.5 meters operating at 140°F. (60°C.) will expand approx. 2″ or 5 cm.

CPVC; Polypropylene; and PVDF at the same temp. would expand approx. 3¼″ or 8 cm.

PLASTIC PIPE AND FITTINGS

Plastic Pressure Piping for hot and cold water suppl is now permitted in FHA-financed rehabilitation projects Plastic pipe enjoys markets in natural gas distribution rural potable water systems, crop irrigation and chemica processing.

Almost 100% of all mobile homes and travel trailer have plastic pipe.

ABBREVIATIONS

ABS—Acrylonitrile-Butadiene-Styrene
PVC—Polyvinyl Chloride
NSF—National Sanitation Foundation
PPI—Plastics Pipe Institute
CPVC—Chlorinated Poly-Vinyl Chloride
PE—Polyethylene Plastic or Resin
PVDF—Polyvinylidene Fluoride

PVC
WATER PRESSURE RATINGS AT 73.4°F. (23°C.) FOR SCHEDULE 40

Pipe Size		PVC-1120-B PVC-1220-B PVC-2120-B CPVC-4120-B		PVC-2110-B		PVC-2112-B		CPVC-4116-B PVC-2116-	
		p.s.i.	kPa	p.s.i.	kPa	p.s.i.	kPa	p.s.i.	kPa
3/8"	(9.5 mm)	620	(4275)	310	(2137)	390	(2689)	500	(3447
1/2"	(12.7 mm)	600	(4137)	300	(2068)	370	(2551)	480	(3310
3/4"	(19 mm)	480	(3310)	240	(1655)	300	(2068)	390	(2689
1"	(25.4 mm)	450	(3103)	220	(1517)	280	(1931)	360	(2482
1¼"	(31.75 mm)	370	(2551)	180	(1241)	230	(1586)	290	(2000
1½"	(38 mm)	330	(2275)	170	(1172)	210	(1448)	260	(1793
2"	(51 mm)	280	(1931)	140	(965)	170	(1172)	220	(1517
2½"	(63.5 mm)	300	(2068)	150	(1034)	190	(1310)	240	(1655
3"	(76 mm)	260	(1793)	130	(896)	160	(1103)	210	(1448
3½"	(89 mm)	240	(1655)	120	(827)	150	(1034)	190	(1310
4"	(101.6 mm)	220	(1517)	110	(758)	140	(965)	180	(1241
5"	(127 mm)	190	(1310)	100	(689)	120	(827)	160	(1103
6"	(152.4 mm)	180	(1241)	90	(621)	110	(758)	140	(965

"ABS"
WATER PRESSURE RATINGS AT 73.4°F. (23°C.) FOR SCHEDULE 40

Nominal Pipe Size		ABS-1210 P.S.I.—kPa		ABS-1316 P.S.I.—kPa		ABS-2112 P.S.I.—kPa	
½"	(12.7 mm)	298	(2055)	476	(3282)	372	(2465)
¾"	(19 mm)	241	(1662)	385	(2655)	305	(2103)
1"	(25.4 mm)	225	(1551)	360	(2482)	282	(1944)
1¼"	(31.75 mm)	184	(1269)	294	(2027)	229	(1579)
1½"	(38 mm)	165	(1138)	264	(1820)	207	(1427)
2"	(51 mm)	139	(958)	222	(1531)	173	(1193)
2½"	(63.5 mm)	152	(1048)	243	(1675)	190	(1310)
3"	(76 mm)	132	(910)	211	(1455)	165	(1138)
4"	(101.6 mm)	111	(765)	177	(1220)	138	(951)
6"	(152.4 mm)	88	(607)	141	(972)	110	(758)

PLASTIC PIPE AND FITTINGS

Recommended Practice for making solvent cemented joints with PVC and ABS pipe and fittings:

Pipe should be cut square, using a fine tooth hand saw and a miter box, or a fine toothed power saw with a suitable guide.

Regular pipe cutters may also be used (a special cutting wheel is available to fit standard cutters). Great care should be taken to remove all burrs and ridges raised at the pipe end.

If ridge is not removed, cement in fitting socket will be scraped from the surface upon insertion, producing a dry joint, and causing probable joint failure.

All burrs should be removed with a knife, file, or abrasive paper.

TEST FIT THE JOINT

Wipe both the outside of the pipe and the socket of the fitting with a clean, dry cloth to remove foreign matter. Mate the two parts witout forcing. Measure and mark the socket depth of the fitting on the outside of the pipe. Do not scratch or damage pipe surface to indicate when the pipe end will be bottomed.

The pipe should enter the fitting at least 1/3 of the socket depth. If the pipe will not enter the socket by that amount, the diameter may be reduced by sanding or filing. Extreme care should be taken not to gouge or flatten the pipe end when reducing the diameter.

PLASTIC PIPE AND FITTINGS

Surface to be joined should be clean and free of moisture before application of the cement.

The outside surface of the pipe (for socket depth) and the mating socket surface shall be cleaned and the gloss removed with the recommended chemical cleaner.

An equally acceptable substitute is to remove the gloss from the mating surfaces (both pipe and socket) with abrasive paper or steel wool.

Wipe off all particles of abrasive and/or PVC before applying cement.

APPLICATION OF CEMENT

Handling Cement: Keep cement can closed and in a shady place when not in use.

Discard the cement when an appreciable change in viscosity takes place, or when a gel condition is indicated by noting that cement will not flow freely from the brush, or when cement appears lumpy and stringy.

The cement should not be thinned. Keep brush immersed in cement between applications.

BRUSH SIZE

The cement is applied with a natural bristle or nylon brush, using ½" (13 mm) brush for nominal pipe sizes ½ inch (13 mm) and less, 1" (25 mm) wide brush for pipe up through 2-inch (51 mm) nominal size, and brush width at least ½ of nominal pipe size for sizes above 2-inch (51 mm), except that for pipe sizes 6-inch (152 mm) and larger a 2½" (64 mm) brush is adequate.

PLASTIC PIPE AND FITTINGS

PVC Solvent Cement is fast drying. Application should be as quickly as possible consistent with good workmanship.

First, apply a full even coat of cement to the pipe surface, to the depth of the fitting socket. Next, apply a uniform, thin coating of cement to the interior of the fitting socket, including the shoulder at the socket bottom.

Recoat the pipe with a second uniform coat of cement, including the cut end of the pipe.

SPECIAL INSTRUCTIONS FOR BELL END PIPE

The procedure as stated thus far may be followed in the case of Bell End Pipe except that great care should be taken not to apply an excess of cement in the Bell Socket, nor should any cement be applied on the bell-to-pipe transition area. This precaution is particularly important for installation of Bell End Pipe with a wall thickness of less than ⅛ inch (3 mm).

ASSEMBLY OF JOINT

Immediately after applying the last coat of cement to the pipe, insert the pipe into the fitting until it bottoms at the fitting shoulder. Turn the pipe, or fitting ¼ turn during assembly (but not after the pipe is bottomed) to evenly distribute the cement.

Assembly should be completed within 20 seconds after the last application of cement.

The pipe should be inserted with a steady even motion. Hammer blows should not be used.

PLASTIC PIPE AND FITTINGS

Until the cement is set in the joint, the pipe may back out of the fitting socket if not held in place for approximately one minute after assembly.

Care should be taken during assembly not to disturb, or apply any force to joints previously made. Fresh joints can be destroyed by early rough handling.

After assembly, wipe excess cement from the pipe at the end of the fitting socket.

A properly made joint will normally show a bead around its entire perimeter.

Any gaps at this point may indicate a defective assembly job, due to insufficient cement, or use of light bodied cement on a gap fit where heavy bodied cement should have been used.

SET TIME

Handle the newly assembled joints carefully until the cement has gone through the set period.

Recommended set time is related to temperature.

Temperature	Set Time
60°F. to 100°F. (15.5 to 37.7°C.)	30 minutes
40°F. to 60°F. (4.4 to 15.5°C)	1 hour
20°F. to 40°F. (−6.6 to 4.4°C.)	2 hours
0°F. to 20°F. (−17.7 to −6.6°C.)	4 hours

PLASTIC PIPE AND FITTINGS

After the set period, the pipe can be carefully placed in prepared ditch. Shade backfill, leaving all joints exposed so that they can be examined during pressure tests.

Test pressure should be 150% of system design pressure and held at this pressure until the system is checked for leaks, or follow requirements of applicable code, whichever is higher:

Note: For most cases, 48 hours is considered to be a safe period for the piping system to be allowed to stand vented to the atmosphere before pressure testing.

Shorter periods may be satisfactory for high air temperatures, small sizes of pipe, quick drying cement, and tight dry fit joints.

PVC and ABS pipe and fittings may be stored either inside or outdoors if they are protected from direct sunlight.

The plastic pipe should be stored in such a manner as to prevent sagging or bending.

Plastic pipe should be supported in horizontal runs as follows:

Nominal Pipe Size	Schedule 40
½" and ¾" (12.7 and 19 mm)	Every 4 ft. (1.22 m)
1" and 1¼" (25 and 32 mm)	Every 4½ ft. (1.37 m)
1½" and 2" (38 and 51 mm)	Every 5 ft. (1.52 m)
3" (76 mm)	Every 6 ft. (1.83 m)
4" (102 mm)	Every 6¼ ft. (1.9 m)
6" (152 mm)	Every 6¾ ft. (2.06 m)

Nominal Pipe Size	Schedule 80
½" and ¾" (12.7 and 19 mm)	Every 5 ft. (1.52 m)
1" and 1¼" (25 and 32 mm)	Every 5½ ft. (1.68 m)
1½" and 2" (38 and 51 mm)	Every 6 ft. (1.83 m)
3" (76 mm)	Every 7 ft. (2.13 m)
4" (102 mm)	Every 7½ ft. (2.29 m)
6" (152 mm)	Every 8½ ft. (2.59 m)

PLASTIC PIPE AND FITTINGS

Note: The industry does not recommend threading ABS or PVC Schedule 40 Plastic Pipe.

Note: A quart can (of the solvent recommended for the type of pipe being joined) is generally sufficient for the average two-bath home.

Note: An ABS stack can be tested within one hour after the last joint is made up.

Note: Common pipe dopes must not be used on threaded joints. Some pipe lubricants contain compounds that may soften the surface which, under compression can set up internal stress corrosion.

If a lubricant is believed necessary, ordinary vaseline or pipe (tape) can be used.

Note: Do not use alcohol or anti-freeze solutions containing alcohol to protect trap-seals from freezing.

Strong saline solutions or magnesium chloride in water (22% by weight) can be used safely.

Glycerol (60% by weight) mixed with water is also recommended.

Note: ABS absorbs heat so slowly that once installed, heat from dishwashers, clothes washers and discharge from similar installations will not cause any problem.

Note: PVC solvent cements are available in two general viscosity categories, namely: light and standard. The light cements are intended for use with pipes and fittings up through 2-inch (51 mm)—N.P.S., and for pipes and fittings where "interference fits"—between the parts to be joined occurs.

Note: Last, but not least, always look for the initials NSF; these initials stand for "National Sanitation Foundation."

The initials NSF will denote approval and standards met as handed down by NSF.

The Foundation is a non-profit, non-commercial organization seeking solutions to all problems involving cleanliness. It is dedicated to the prevention of illness, the promotion of health and the enrichment of the quality of American living through the improvement of the physical, biological and social environment in which we live today.

The NSF seal on DWV and potable water plastic pipe and fittings, means compliance with Foundation policies and standards.

The NSF standards, research and education programs are designed to benefit all parties: the manufacturer, the regulatory officials, building industries, the product specifier, installer, the ultimate user and, maybe most important of all, aid in providing protection to public health.

Plastic Pipe and Fittings

PLASTICS PIPE INSTITUTE Presents....

"CEMENTING of PVC PRESSURE PIPE"

Plastic Pipe and Fittings

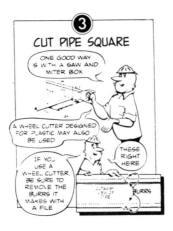

Plastic Pipe and Fittings

Plastic Pipe and Fittings

PLASTIC PIPE AND FITTINGS

PLASTIC PIPE AND FITTINGS

PLASTIC PIPE AND FITTINGS

PLASTIC PIPE AND FITTINGS

Lead and Oakum Joints

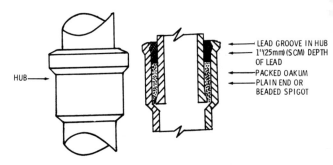

- LEAD GROOVE IN HUB
- 1"(25mm)(5 CM) DEPTH OF LEAD
- PACKED OAKUM
- PLAIN END OR BEADED SPIGOT

HUB

LEAD AND OAKUM JOINTS

Tools Needed: "Up to 6" or 152 mm joints"

1. 12 oz. or .340 kilogram ball peen hammer
2. Caulking iron or irons
3. Packing iron
4. Yarning iron
5. Joint runner (for pouring lead in horizontal joints)
6. Wood chisel (steel handle) for cutting "lead gate" created by use of joint runner.
7. Chain snap cutters or ratchet cutters (for cutting pipe).

Note: A cold chisel and 16 oz. (.453 kg.) ball peen hammer can be used to cut pipe. Place a 2" × 4" or 5 × 10 cm piece of wood directly under cut mark; allow one end of pipe to touch ground, with one foot placed near desired cut—hold pipe solid against 2" × 4".

Begin marking pipe with light hammer blows. When entire circumference of pipe is marked with cold chisel indentation marks, BEGIN USING HEAVIER BLOWS.

8. Six foor rule or metric stick.

Lead and Oakum Joints

After proper amount of Oakum is placed in joint, leaving one inch (25 mm) for lead (joints up to 6-inch (152 mm)) lead is then poured. Joint now ready for caulking to make it water and air tight.

Note: Before dipping "ladle" in molten lead, be sure "ladle" is dry and free from moisture. Warm it over lead pot while lead is being heated.

Moisture on a "ladle" will form steam when dipped into molten lead — causing an explosion.

Caulking joint:
   "Horizontal Joints": Caulk inside first.
   "Vertical Joints": Outside first is preferred.

Note: When "Caulking" — use moderate hammer blows, each position of the "caulking iron" should slightly overlap the previous position.

Note: The "Lead and Oakum Joint" provides a waterproof joint. Strong, flexible, root proof, water and air tight.

LEAD AND OAKUM INFORMATION

Size of Pipe and Fitting	Lead Ring Depth	White Oakum Ounces	Brown Oakum Ounces	Lead SV — XH
2" Joint	1"	1½ oz.	1¾ oz.	1¼ lbs
3"	1	1¾	2½	1¾
4"	1	2¼	3	2¼
5"	1	2½	3¼	2¾
6"	1	2¾	3½	3
8"	1¼	5¼	7	6
10"	1¼	6½	8½	8
12"	1¼	7½	9¾	10¾
15"	1½	11½	15	17¾

Note: Approx. 8 lbs. (3.62 kg) "Brown Oakum" is used per 100 lbs. (45.3 kg) of lead.

6 lbs. (2.7 kg) white oakum is used per 100 lbs. (45.3 kg) of lead.

Note: "Caulking Lead" in cast-iron bell and spigot water mains should be 2 inches (51 mm) deep.

METRIC

Size of Pipe and Fitting	Lead Ring Depth	White Oakum Grams	Brown Oakum Grams	Lead SV-XH Grams
51 mm	25 mm	43 g	50 g	567 g
76 mm	25 mm	50 g	71 g	794 g
102 mm	25 mm	64 g	85 g	1020 g
127 mm	25 mm	71 g	92 g	1247 g
152 mm	25 mm	78 g	99 g	1360 g
203 mm	32 mm	149 g	198 g	2721 g
254 mm	32 mm	184 g	241 g	3629 g
305 mm	32 mm	213 g	276 g	4876 g
381 mm	38 mm	326 g	425 g	8051 g

SOLAR SYSTEM WATER HEATERS ILLUSTRATED

I wish to thank the A. O. Smith Corporation, Consumer Products Division, P.O. Box 28, Kankakee, Illinois 60901, for providing the illustrations and data on their new—"Conservationist Solar System Water Heater."

America's dwindling supply of energy is becoming more critical as each new year passes. And with the cost of electricity and gas rising towards astronomical heights, it may be wise for all of us to take a more closer look at SOLAR ENERGY; or at least become familiar with a system which, in my opinion, is here to stay.

The Conservationist Solar System Water Heater—illustrated on the following pages—can be tailored to any area and prevailing condition. A. O. Smith Solar Systems can be designed for use with existing hot water heaters.

THE CONSERVATIONIST SOLAR SYSTEM WATER HEATER

SOLAR SYSTEM WATER HEATING ILLUSTRATED

Here's how the A. O. Smith CONSERVATIONIST solar system works:

The hot sun rays are absorbed by roof-mounted collector panels to heat special antifreeze fluid that is circulating through integral copper channels.

A. O. Smith utilizes a closed-loop system for transfer of heated solution and return. Propylene glycol eliminates any worries of freezing.

Heater-mounted differential controller has modulating output to collect maximum amount of available heat from collector panels, even on cloudy days. Pump is adjustable for flow with a restrictor that makes the A. O. Smith CONSERVATIONIST solar system flexible for various installations.

Diaphragm expansion tank is provided on top of heater to handle expansion of heat transfer fluid in closed-loop circulating line.

Two high density magnesium anodes protect tank against corrosion.

Three-inch double efficiency blanket of high density insulation surrounds tank to keep in more heat.

Tank is isolated from the jacket to prevent conduction heat loss.

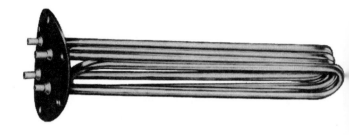

EXCLUSIVE CORONA™ HEAT EXCHANGER

The A. O. Smith Corona™ heat exchanger is immersed in the tank to assure direct transfer of the heat. Ordinary exchangers are less efficient with a wrap-around-tank method. The Corona™ has a double wall of copper for safety and is electrically isolated from the tank and external piping for positive protection against corrosion.

The Corona™ heat exchanger is used in Conservationist™ solar water heaters—models SUN-82, 100 and 120 gallon, or 378.5 and 454.25 liter sizes.

A. O. Smith Phoenix™ screw-in immersion element has two-way protection: A sheathing of Iron-Base Superalloy provides excellent protection against burn-out, oxidation and scaling. The ceramic terminal block won't melt like ordinary plastic blocks. The Phoenix™ elements provide "back-up" heating, as needed.

Solar System Water Heaters

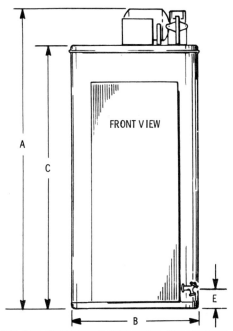

SOLAR SYSTEM WATER HEATER

Model No.	"A" Inches	Millimeters	"B" Inches	Millimeters	"C" Inches	Millimeters	"E" Inches	Millimeters
Sun-82	56	1422	28	711	48	1219	4¼	108
Sun-100	65⅞	1673	28	711	57⅞	1470	4¼	108
Sun-120	69	1753	30	762	61	1549	5¾	146

Model No.	Capacity U.S. Gals.	Liters	Approx. Ship. Wt. Pounds	Kilograms
Sun-82	82	310.3	235	106.5
Sun-100	100	378.5	250	113.4
Sun-120	120	454.25	340	154.2

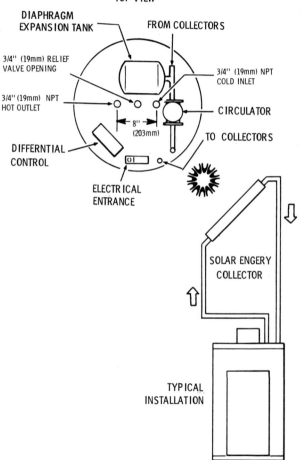

TIPS FOR THE PLUMBER—BEGINNER

When a measurement is taken from a blueprint it should be checked from both ends of the building to assure accuracy.

When running pipe in ceiling where it must be hung, you should always transfer your work to the floor—where possible; transferring sleeve locations from floor above to floor below. Also make use of columns and walls to square off your work.

Once points to be reached are established and you now know in which direction you are heading, and where you are coming from, it will now become clear what fittings will work. Cuts can then be determined, and even location of hangers, if inserts were not provided in the building construction.

The next step is transfering these hanger locations to the ceiling by way of your plumb bob.

268   Tips for the Plumber

When running hot and cold water headers to a number of fixtures in the same area, especially a battery of lavatories, run one header low—the other header high. Example:

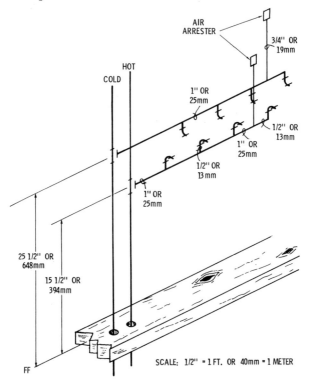

Tips for the Plumber

"COMMON TERMS TO REMEMBER"

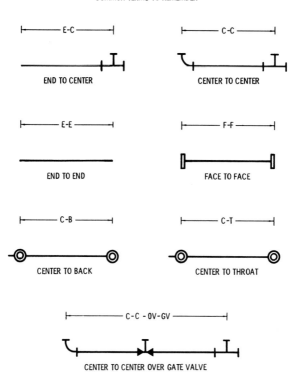

When installing a water closet (floor mounted) to closet floor flange, be it plastic, brass, or cast iron, be careful as you tighten closet nuts. Tighten both sides evenly, and tighten just enough that bowl doesn't rock. As soon as nuts are drawn up snug—sit down on bowl to settle wax seal in place, then snug up a little more. Remember, too tightly will crack bowl!

On water closet flush valves and lavatory supply tubes, tighten from the top down. This will avoid possible leaks and a waste of time.

When P-traps from lavatories must be soldered into a lead or copper waste pipe, secure all piping in place so that the soldering of trap outlet is the last operation.

Lastly, acquire the habit of using two wrenches when tightening or loosening pipe, and you'll avoid lots of unnecessary problems.

SLOAN ROYAL FLUSH VALVE ILLUSTRATED

I wish to thank the Sloan Valve Company, 10500 Seymour Avenue, Franklin Park, Ill. 60131, for providing the illustrations and data on the Sloan Royal Flush Valve.

Incidentally, all Sloan Flush Valves manufactured since 1906 can be repaired. It is recommended that when service is required, all inside parts be replaced so as to restore the flush valve to like-new condition. To do so, order parts in kit form.

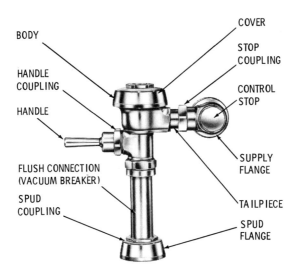

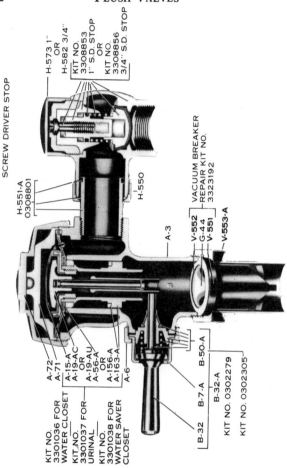

Sloan Royal (diaphragm-type) flush valve.

FLUSH VALVES 273

SLOAN ROYAL (DIAPHRAGM TYPE) FLUSH VALVE
SINCE MID-YEAR 1971

The Repair Kits and Parts listed are designed to service All Sloan [Diaphragm Type] Exposed and Concealed Flush Valves. Each item has been identified by a Part Number along with a corresponding Code Number. To expedite your replacement requirements, order by Code Number.

NEW RETRO WATER SAVER KIT FOR SYPHON JET WATER CLOSETS CODE NO. 3301038

SEE CONTROL STOP REPAIR KITS

For use on Pedal Type Flush Valves only.

V-500-AA Vacuum Breaker (Replaces V-100-AA)

NOTE: For information on V-500-A, V-500-AA Vacuum Breakers - Exposed & Concealed Flush Connections - Tailpieces longer than regular and items not shown consult your local Plumbing Wholesaler.

FLUSH VALVES

PARTS LIST

1. 0301172 *A-72 CP Cover
2. 0301168 A-71 Inside Cover
3. 3301058 A-19-AC Relief Valve (Closet) - 12 per pkg.
   3301059 A-19-AU Relief Valve (Urinal) - 12 per pkg.
4. 3301111 A-15-A Disc - 12 per pkg.
   0301112 A-15-A Disc (Hot Water)
5. 3301188 A-156-A Diaphragm w/A-29 - 12 per pkg.
   0301190 A-156-A Diaphragm (Hot Water)
6. 3301236 A-163-A Guide - 12 per pkg.
7. 3301036 Inside Parts Kit for Closets, Service Sinks, Blowout and Siphon Jet Urinals
8. 3301037 Inside Parts Kit for Washdown Urinals
8A. 3301038 Retro Water Saver Kit-delivers 3-1/2 gal.
9. 3301189 A-156-AA Closet/Urinal Washer Set - 6 per pkg.
10. 3302297 B-39 Seal - 12 per pkg.
11. 3302279 B-32-A CP Handle Assem. - 6 per pkg.
12. 0301082 *A-6 CP Handle Coupling
13. 0302109 B-7-A CP Socket
14. 3302274 B-32 CP Grip - 12 per pkg.
15. 3302305 B-50-A Handle Repair Kit - 6 per pkg.
16. 0303351 C-42-A 3" CP Push Button Assem.
17. 3303347 3" CP Push Button Replacement Kit
18. 3303396 C-64-A 3" Push Button Repair Kit
19. 0306125 F-5-A 3/4" CP Spud Coupling Assem.
    0306132 F-5-A 1" CP Spud Coupling Assem.
    0306140 F-5-A 1-1/4" CP Spud Coupling Assem.
    0306146 F-5-A 1-1/2" CP Spud Coupling Assem.
20. 0306052 *F-2-A 3/4" CP Outlet Coupling Assem.
    0306077 F-2-A 1" CP Outlet Coupling Assem.
    0306092 *F-2-A 1-1/2" CP Outlet Coupling Assem. w/S-30
    0306060 *F-2-A 1-1/4" CP Outlet Coupling Assem.
    0306093 *F-2-A 1-1/2" CP Outlet Coupling Assem.
21. 3323192 V-500-A & V-500-AA Vacuum Breaker Repair Kit
22. 0308676 H-550 CP Stop Coupling
23. 0308801 *H-551-A CP Adj. Tail 2-1/16" Long
24. 0308757 H-600-A 1" SD Bak-Chek CP Control Stop
    0308676 H-600-A 3/4" SD Bak-Chek CP Control Stop
    0308881 *H-600-A 1" WH Bak-Chek CP Control Stop
    0308889 *H-600-A 3/4" WH Bak-Chek CP Control Stop
25. 0308063 *H-6 CP Stop Coupling
26. 0308026 *H-5 CP Ground Joint Tail 1-3/4" Long
27. 0308884 H-650-AG 1" SD Bak-Chek CP Control Stop
    0308882 *H-650-AG 1" WH Bak-Chek CP Control Stop

* Items also available in Rough Brass - Consult Local Plumbing Wholesaler for Code Number.

Parts list for flush valve on page 273.

FLUSH VALVES
SLOAN ROYAL (DIAPHRAGM TYPE) FLUSH VALVE
PRIOR TO MID-YEAR 1971

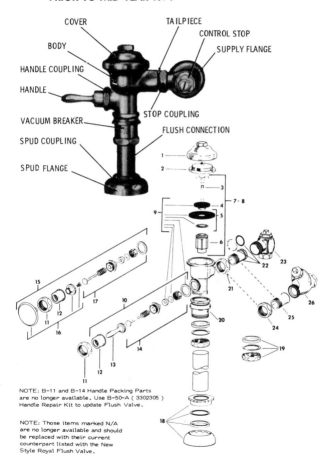

NOTE: B-11 and B-14 Handle Packing Parts are no longer available. Use B-50-A (3302305) Handle Repair Kit to update Flush Valve.

NOTE: Those items marked N/A are no longer available and should be replaced with their current counterpart listed with the New Style Royal Flush Valve.

PARTS LIST

1. Cover CP N/A use 0301172 and 0301168
2. Inside Brass Cover N/A use 0301168 and 0301172
3. A-19-A Brass Relief Valve N/A use 3301058 or 3301059
4. 3301111 A-15-A Disc - 12 per pkg.
   0301112 A-15-A Disc (Hot Water)
5. 3301170 A-56-A Diaphragm w/A-29 - 12 per pkg.
6. Brass Guide N/A use 3301236 - NOTE: 3301236 A-163-A Guide replaces all previous Guides.
7. Inside Parts N/A see item No. 7 listed with new style valve- Repair Kit replaces all previous inside parts.
8. Inside Parts N/A see item No. 8 listed with new style valve- Repair Kit replaces all previous inside parts.
9. 3301176 A-56-AA Washer Set - 6 per pkg.
10. B-32-A CP Handle Assem. N/A use 3302279
11. A-6 CP Handle Coupling N/A use 0301082
12. B-7 CP Socket N/A use 0302109
13. B-32 CP Grip N/A use 3302274
14. Handle Repair Kit N/A use 3302305
15. C-42-A 3" CP Push Button Assem. N/A use 0303351
16. 3303347   3" CP Push Button Replacement Kit
17. 3" Push Button Repair Kit N/A - use 3303396
18. Spud Coupling Assem. CP N/A - see item No. 19 listed with new style valve.
19. Outlet Coupling Assem. CP N/A - see item No. 20 listed with new style valve.
20. V-100-A & V-100-AA Vacuum Breaker N/A consult local Plumbing Wholesaler for proper V-500-A or V-500-AA Vacuum Breaker replacement.
21. H-550 CP Coupling N/A -use 0308676
22. *0308801 H-551-A CP Adj. Tail 2-1/16" long
23. H-540-A Series Control Stops N/A - see Control Stop Repair Kits or item No. 24 listed with new style valve for complete replacement.
24. H-6 CP Stop Coupling N/A use 0308063
25. *0308026 H-5 CP Ground Joint Tail 1-1/4" long
26. H-545-AG Series Control Stops N/A - see Control Stop Repair Kits or item No. 27 listed with new style valve for complete replacement.

Parts list for flush valve on page 275.

WALL-HUNG WATER COOLER ILLUSTRATED

I wish to thank the EBCO Manufacturing Company, 265 North Hamilton Road, Columbus, Ohio 43213, manufacturers of Oasis Water Coolers, for providing the illustrations and data contained in the following section.

EBCO is one of the largest manufacturers of electric water coolers—dating back to 1924.

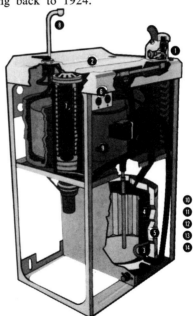

Wall-hung water cooler.

1. Dial-A-Drink Bubbler—assures smooth, even flow of water under pressures from 20 to 125 PSI (138 to 862 kPa). Exclusive design minimizes wear to assure long lasting, trouble-free operation.

2. Stainless Steel Top—of polished 18-8 stainless resists rust, corrosion, stains. Anti-splash ridge and integral drain direct and dispose of water.

3. Red Brass Cooling Tank—offers maximum cooling efficiency, reduces starts of compressor. 85-15 red brass storage tank (vented) has internal heat transfer surface and external refrigeration coil bonded to tank by immersion in pure molten tin.

4. Copper Cooling Coils—around storage tank insure maximum cooling efficiency. Double-wall separation of refrigerant and drinking water conforms to all sanitary codes.

5. Insulating Jacket—of expanded polystyrene foam maintains cold water temp. on all models.

6. Adjustable Thermostat—is tamper-proof. Remote sensing bulb located in cooling tank provides accurate control of cold water temperature.

7. Cost cutting pre-cooler—(on larger capacity models) nearly doubles capacity without extra operating cost by cooling incoming water with cold waste water.

8. Hot water availability—on Hot 'N Cold models. Hot tank heats and serves up to 45 cups of piping hot water per hour. Hot water system is atmospheric vented and fiberglass insulated.

9. Refrigeration system—is maintenance free. Compressor and motor are hermetically sealed, lubricated for life and virtually leakproof.

10. Durable, handsome cabinet finish—includes Vinyl laminated-on steel on front and side panels, which provide a good-looking, durable, scuff-resistant finish.

11. ARI Certified Performance—means cooling capacity of an Oasis water cooler is rated and certified in accordance with Air Conditioning and Refrigeration Institute (ARI) Standard 1010-73 (A.N.S.I. Standard A112.-11.1-1973): Gallons per hour of 50°F. (10°C.) drinking water with inlet temp. of 80°F. (27°C.), and room temp. of 90°F. (32°C.).

12. Ease of installation — All On-A-Wall, Semi-Recessed and Simulated Semi-Recessed models have removable front and side panels. Provide extra work space for plumbing and electrical installations, servicing and routine maintenance.

13. Thoroughly tested—Each Oasis water cooler is subjected to continuous assembly-line tests and a complete capacity performance test. Each Ebco Water Cooler carries a limited five-year warranty.

WALL-HUNG WATER COOLER
Dimensional Drawing

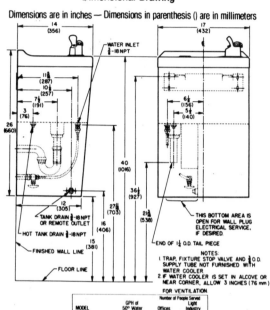

MODEL	GPH of 50° Water	Number of People Served Offices	Light Industry
ODP16M	15.7	188	100
ODP13M	13.0	156	91
*ODP13M-60	13.0	156	91
ODP7M	7.0	84	49
ODP7MH	7.0	84	49
ODP5M	5.0	60	35
ODPM	Non-Refrigerated Fountain		
ODP15MW	Water cooled. See page 17		

NOTES:

Rough in above water cooler at:

Waste: 22½″ or 572 mm, 5½″ (140 mm) left of center line.

Water: 17½″ or 445 mm, 6½″ (165 mm) left of center line.

Water Supply Pipe: ½″ N.P.S. or 13 mm—reduced to ⅜″ O.D. (10 mm). Waste Piping is 1¼″ I.D. or 32 mm.

PLUMBING TOOLS ILLUSTRATED

I wish to thank the Ridge Tool Company, 400 Clark Street, Elyria, Ohio 44035, manufacturer of Ridgid pre-tested work saver tools for providing the tool illustrations in this section.

Ball peen hammer.

Four Sizes: 2" through 8" (51-203 mm)

Here is a wrench that lets one man do the work of two. Short handle makes it easy to get at frozen joints even in tight quarters.

RIDGID Straight Pipe Wrenches

Ten Sizes, 6" through 60" or 152 thru 1524 mm

These, the world's most popular pipe wrenches, are known for the brutal punishment they can take because of their extra built-in toughness. Before shipment every wrench is work tested. Housing is replaced free if it ever breaks or distorts. Replaceable jaws are made of hardened alloy steel. Full-floating hook jaw assures instant grip . . . easy release. Spring suspension eliminates chance that jaws could jam or lock on pipe. Handy pipe scale and large, easy-to-spin adjusting nut give fast, one-hand setting to pipe size. Comfort-grip, malleable iron I-beam handle has convenient hang-up hole.

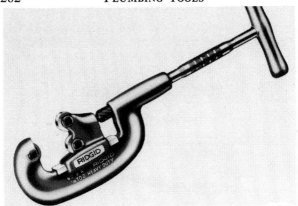

Heavy-duty pipe cutter; cuts pipe ⅛" thru 2" (3 thru 51 mm).

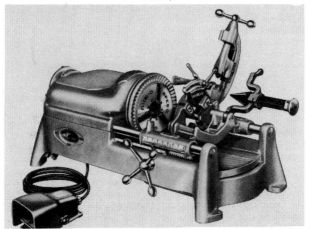

Pipe and bolt threading machine, ⅛" thru 2" (3 thru 51 mm) pipe. This machine cuts, threads, reams and oils.

Plumbing Tools 283

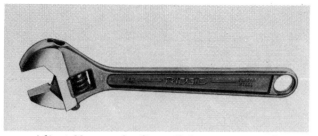

Adjustable wrench, often called crescent wrench.

RIDGID No. 342 Internal Wrench. Holds closet spuds and bath, basin and sink strainers through 2″ (51 mm). Also handy when installing or extracting 1″ through 2″ (25 thru 51 mm) nipples without damage to threads.

RIDGID Basin Wrenches. (A) RIDGID No. 1010 Basin wrench has solid 10″ (254 mm) shank. (B) RIDGID Nos. 1017 and 1019 Basin wrenches have telescopic shanks for lengths from 10″ through 17″ (254 thru 431 mm).

PLUMBING TOOLS

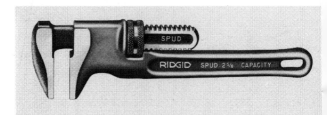

Spud wrench; capacity 2⅝" (66.5 mm).

Strap wrench—⅛" thru 2" (3 thru 51 mm).

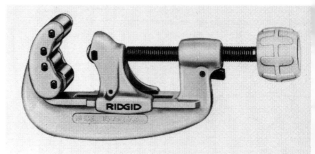

Quick-acting tubing cutter; ¼" thru 2⅝" (6 thru 66.5 mm).

PLUMBING TOOLS 285

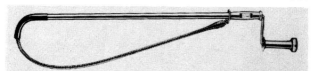

Closet auger, used for water closet and urinal stoppage.

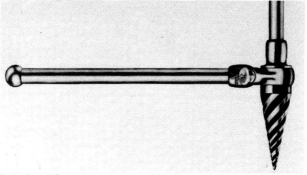

Spiral ratchet pipe reamer; ⅛" thru 2" (3 thru 51 mm).

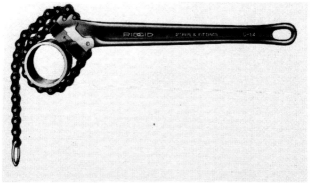

Heavy-duty 2" (51 mm) chain wrench.

Plumbing Tools

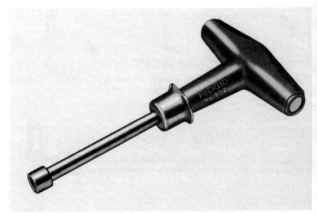

Torque wrench for cast-iron and no-hub soil pipe. Pre-set for 60 inch lbs. (67.9 N·m) of torque.

Flaring tool—will flare tubing size from 3/16" (4.8 mm) 1/4" (6.4 mm), 5/16" (7.9 mm), 3/8" (9.5 mm), 7/16" (11.1 mm), 1/2" (12.7 mm) and 5/8" (15.9 mm).

PLUMBING TOOLS 287

Ratchet cutter for cast-iron pipe—2" thru 6" (51 thru 152 mm).

Soil pipe assembly tool—2" thru 8" (51 thru 203 mm).

Straight pipe welding vise.

Angle pipe welding vise.